D1545806

SOLDERING PROCESSES AND EQUIPMENT

SOLDERING PROCESSES AND EQUIPMENT

Edited by

MICHAEL G. PECHT

CALCE Electronic Packaging Research Center
University of Maryland
College Park, Maryland

A Wiley–Interscience Publication

JOHN WILEY & SONS, INC.

New York / Chichester / Brisbane / Toronto / Singapore

This text is printed on acid-free paper.

Library of Congress Cataloging in Publication Data:

Pecht, Michael.

Soldering processes and equipment / Michael G. Pecht.

p. cm.

Includes bibliographical references and index.

ISBN 0-471-59167-X (alk. paper)

1. Electronic packaging–Materials. 2. Solders and soldering.
3. Surface mount technology. I. Title

TK7870.15.P43 1993

621.381'046–dc20 92-33770

CIP

Printed in the United States of America

10 9 8 7 6 5 4 3 2 1

To my wife, Judy,
daughter, Joan,
sons, Jefferson and Andrew,
my parents, George and Dorothy,
and grandma Shih

Contributors

Louis A. Abbagnaro has a B.S. in Electrical Engineering from Yale and an M.S. in Electrical and Mechanical Engineering from the University of Bridgeport. He worked as director of audio R&D at CBS Tech Center from 1964 to 1980 and was vice president of engineering for CBS Electronics and Toys until 1986. He is currently vice president and general manager of Pace, Inc. in Laurel, MD. Mr. Abbagnaro is a fellow of the Audio Engineering Society and has been a member of various technical committees of the Audio Engineering Society, EIA, IPC and ISO. He has authored several papers on audio, acoustics, engineering design and rework and repair.

Sarvotham M. Bhandarkar obtained his Ph.D. in Mechanical Engineering from the University of Maryland in 1992. His dissertation research was on thermomechanical fatigue analysis of interconnections in multilayered PWBs. His expertise is in structural and thermal analysis, mechanics of composite materials, materials characterization, and numerical optimization techniques, and he has several publications in these areas. As a member of the CALCE research staff, he is currently involved in investigating structural failure mechanisms in the high-density interconnect (HDI) technologies used in multichip modules.

David A. Curtis is widely recognized as an authority in the field of electronics manufacturing whose consulting assignments have taken him to Europe, Asia, and Australia. His clients range from small family-owned businesses to companies high on the *Fortune 500* list, and he has worked with many U.S. and foreign government agencies. He assists senior executives of client companies with the evaluation of new or improved technologies, and with their implementation in new or existing factories.

Prior to founding the consulting company David A. Curtis & Associates, Inc. in 1981, he was a vice-president of Booz, Allen & Hamilton, Inc., in the firm's Technology Management Group. He was also vice president in the Electronic Systems Section of Arthur D. Little, Inc.; a director of Cambridge Consultants, Ltd., in the United Kingdom; a lecturer in applied physics at the University of Durham (U.K.); and a NATO research fellow in the Kamerlingh Onnes Laboratorium of the University of Leiden (The Netherlands).

He is a senior member of the Society of Manufacturing Engineers and of the Institute of Electrical and Electronics Engineers. He is chairman-elect of the Association for Electronics Manufacturing of the SME and vice president for publications of the IEEE's Engineering Management Society.

Dr. Curtis is a frequent speaker and author. His latest book, *Making a Success in Manufacturing* will be published by McGraw-Hill in 1993. He is listed in *Who's Who in the East.*

Anupam Malhotra received his B.E. in Mechanical Engineering from Panjab University, India, and is currently pursuing an M.S. in Mechanical Engineering at the University of Maryland, College Park. His fields of interest include soldering technology and reliability program development based on the physics-of-failure of electronic components, devices, and processes. He is currently working on the restructuring of MIL-STD-785, Reliability Program for Systems and Equipment, Development and Production, into a physics-of-failure based com-military document and on the implementation of the revised document as a national IEEE standard.

Michael G. Pecht has a B.S. in Acoustics, an M.S. in Electrical Engineering, and an M.S. and Ph.D. in Engineering Mechanics from the University of Wisconsin. He is a Professional Engineer and IEEE Fellow, and serves on the board of advisors for the Society of Manufacturing Engineers Electronics Division, the *Journal of Electronics Manufacturing*, and the *Journal of Concurrent Engineering.* He is the director of the CALCE Electronic Packaging Research Center at the University of Maryland. He is also the chief editor of the *IEEE Transactions on Reliability*, and has recently edited a book titled *Handbook of Electronic Package Design*, published by Marcel Dekker (1991).

Shailendra Verma received his M.S. in Mechanical Engineering from the University of Maryland, College Park. His fields of interest include modelling of failure mechanisms in solder joints using non-linear finite-element techniques, integrating reliability tools for product validation, and process verification and stress screening. He received his B.Tech. from the Indian Institute of Technology, Kharagpur, in 1990.

Contents

Preface

In the last three decades, soldering technology has developed from an art into a high-technology science — an evolution chiefly influenced by the pressure placed on the soldering industry by advances in microelectronics. This book addresses the major facets of modern soldering technology and the science behind that technology.

Chapter 1 defines soldering and solder joints and describes the various steps involved in soldering different types of printed circuit and wiring boards. The processes of soldering that can be automated are also discussed. The uses of solder for purposes other than for connecting components to a printed wiring board are presented, as are techniques emerging as viable alternatives to soldering.

Chapter 2 presents an overview of soldering in terms of materials. Various alloys, fluxes, and pastes used in soldering are described, and issues related to intermetallics and contaminants are examined. Flux materials and their mechanical, electrical, thermal, and physical properties are presented.

Chapter 3 discusses wave soldering and flux, preheat, solder waves, solder pots, pumps, nozzles, and conveyers. The dynamic concept of the solder wave is discussed and analyzed. State-of-the-art wave soldering methods, such as dual wave soldering, ultrasonic soldering, and controlled atmosphere soldering, are also described.

Chapter 4 focusses on reflow soldering. Dispenser technology and the various methods of depositing solder paste and adhesive on the printed board are presented, along with screen printing and pin-transfer methods. Infrared, hot bar, laser, hot plate, and vapor phase reflow soldering are discussed and their advantages and disadvantages are identified. The state-of-the-art technologies of microflame and optical fiber reflow soldering, along with special equipment being manufactured for soldering fine-pitch components, are examined.

Chapter 5 describes cleaning processes and the important properties of cleaning agents and cleaners. The Montreal Protocol, requiring the phased elimination of CFCs from the environment, is discussed. Semi-aqueous and aqueous cleaning systems and the upcoming no-clean soldering processes using specially formulated fluxes are also presented.

Chapter 6 focusses on solder joint reliability and quality, using a physics-of-failure approach. Intermetallic compounds, which form whenever two different metals are soldered together, are addressed. Test and evaluation procedures adopted by the soldering industry to determine solder joint reliability (including non-destructive evaluation methods and acceleration testing techniques) are examined. Statistical process control and screening methods are also considered.

Chapter 7 emphasizes the importance of rework in the soldering industry. Although soldering has evolved into a high-tech science, the industry is still far from its goal of six-sigma manufacturing; therefore, proper rework and repair procedures must often be used to lower manufacturing costs. Repair and rework operations performed on through-hole, as well as surface-mount, devices are described. Guidelines on component selection and assembly techniques are also given. The latest equipment in the repair industry is described for both through-hole and surface-mount devices. Other rework tasks, including track repair, electroplating, and cleaning, are also described, as is the modern modular rework station.

Appendix A lists various solder equipment manufacturers. Appendix B provides a glossary of soldering terms.

I would like to thank many people for their help and contribution to this book. Without their cooperation, this effort would never have gotten past its infancy. In particular, sincere appreciation is given to Carol Lalley of E-Systems; Bob Doetzer of Methods Automation; Arlene Kormis, Dave Robertson, and Eric Seigel of Pace, Inc.; Tom O'Connor of T & M Sales Co.; Abe Glieberman of AM&P, Inc.; Werner Engelmaier of Engelmaier Associates; and Abhijit Dasgupta, Richard Cogan, Tom Riddle, Ashok Prabhu, Joanyang Chang and Amal Hamad of the CALCE Electronic Packaging Research Center at the University of Maryland. A special thanks is also due to the various industry personnel who provided informative articles and numerous photographs for the book. I would also like to thank Frank Cerra and Bob Hilbert of John Wiley and Sons for publishing the book in a timely manner.

Michael Pecht

SOLDERING PROCESSES AND EQUIPMENT

Chapter 1

Introduction

Soldering is not a new technology; it has existed since the early bronze age, and the Romans used an alloy of tin and lead for sealing their water pipes. Soldering of electronics started after the introduction of the vacuum tube by Lee Deforest in 1907. However, it was only after the automation of the printed wiring board assembly process that soldering became the fundamental and most common method for electrical connections. Since then, many new soldering processes have been developed to improve the quality and reliability of the solder joints on a printed circuit board. The equipment used in soldering now includes fully automated machinery with computer vision for precision and accuracy.

1.1 Reasons to solder

Metals may be joined together by many techniques, including mechanical fastening, using nuts, bolts or rivets; adhesive bonding; and metallurgical welding, brazing and soldering. Of these techniques, metallurgical fastening provides the best electrical and thermal conductive paths and can be used to form environmental seals around the joining metals. Among metallurgical techniques, soldering differs from welding in that only the filler metal (solder) is melted to form the bond. In welding, the metals themselves are fused together at high temperatures to form the bond. Soft soldering requires low temperatures (up to 400°C) to melt the filler metal (solder) to create the joint and is the only metallurgical technique with a temperature range practical for joining electronic assemblies.

The relatively low cost of soldering alloys and the ease of manufacturing them has made soldering an economically viable joining process for mass production in electronics. As a metal-joining technique for electronic assemblies, solder has many advantages, including low-temperature manufacturing, good electrical and thermal conductive paths, support to withstand thermomechanical and vibrational loads, and easy automation for mass production of a wide variety of assemblies.

1.2 Definition of a solder joint

In the past, soldering processes principally involved soldering of through-hole components to a printed wiring board. This process required inserting the component leads through the plated through-holes prior to soldering the leads. Current technologies use both through-hole and surface-mount devices (SMDs) to achieve dense packaging on the board. The SMDs are leaded or leadless devices placed and soldered onto the board without inserting the leads through holes. In this case, the solder joint provides the mechanical attachment of the device, as well as electrical, and in some cases thermal, continuity.

The geometry of the solder joint depends on the type of package (component) being soldered to the board. Packages are broadly classified into through-hole components and surface-mount devices (SMDs), and into several categories, dependent on the lead types. In the case of through-hole packages, the leads are either soldered straight through or are clinched (bent over) before the soldering operation (Figure 1.1), leading to different fillet shapes for the solder joint. The mechanical attachment is partially provided by inserting the leads through the hole. The solder joint completes the mechanical attachment and fills in the gaps to provide electrical and thermal continuity between the lead and the through-hole walls.

For surface-mount packages, the solder alone is the mechanical attachment bearing the stresses resulting from any vibrational and thermomechanical loads (Figure 1.2). During assembly, surface-mount packages must be held in place on the solder pads from the time of placement to the time of actual soldering. The complex material and processing constraints on surface-mount packages have led to the use of specialized materials for package placement, such as solder pastes and adhesives, and to the development of fatigue-resistant solder alloys and specialized automated soldering processes.

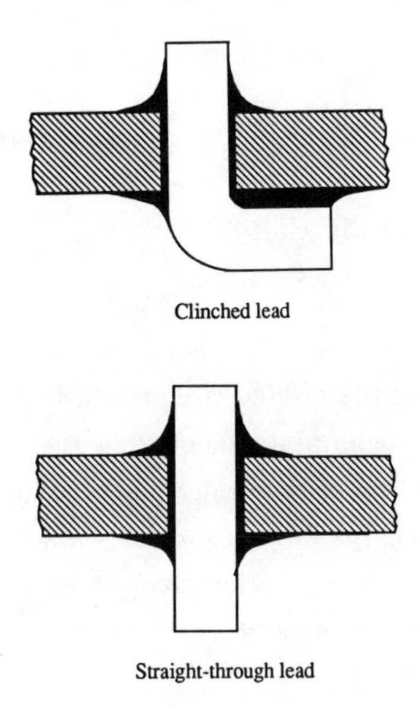

Figure 1.1: Solder joint geometries for through-hole packages

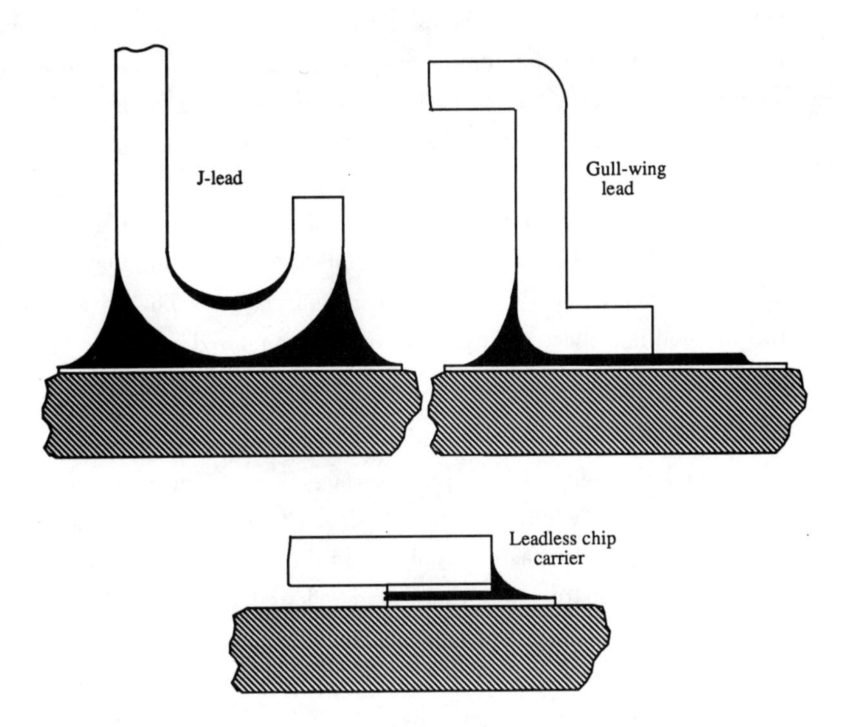

Figure 1.2: Solder joint geometries for surface-mount packages

1.3 Soldering processes and classification

Soldering processes vary with the type of components (surface-mount, through-hole, or both) to be placed and the level of integration required. Electronic board assemblies are usually classified as single-sided, double-sided, mixed double-sided, or mixed packages.

- **Single-sided assemblies:** In this assembly, the components are placed on only one side of the printed circuit board. Soldering begins with the application of the solder paste, followed by the placement of components, preheating and reflow of the solder paste, cleaning of the assembly, inspection, and touch-up. Rework processes are done if needed. The main application of such a system is low component-density boards. Compared to through-hole mount processes, surface-mount single sided boards save up to 40% board real estate.

- **Double-sided assemblies:** High-density boards usually have components and devices attached on both sides. After the solder paste is applied, an adhesive paste holds down components when the board is reversed to attach more components. Components are placed over the adhesive, the board is preheated and the solder paste is then reflowed. The board is overturned after cleaning and solder paste is applied on the secondary side. After the components are placed on it, the board goes through another cycle of preheating and reflowing before being inspected and touched-up. Testing and rework follow. Such assemblies are usually found where the PCB has at least four layers. The advantage of this method is a space saving of 60 to 70%, compared with through-hole processes.

- **Mixed double-sided assemblies:** Mixed assemblies use both surface-mount and through-hole devices. Surface-mount devices are attached with solder paste, while through-hole devices and components are wave soldered. After solder paste is applied and the surface-mount components are placed, the board is preheated and reflowed. The assembly is then cleaned and the board is overturned. Pick-and-place machines automatically insert through-hole devices into their respective places on the board. The board is overturned again, and adhesive is applied on the primary side, followed by placing the components and curing the adhesive. The board is overturned a third time, and the through- hole components are manually inserted into the secondary side and wave soldered. The assembly is then cleaned, inspected, and touched-up. Testing and rework are done if necessary. This method provides the best use of the printed circuit board's real estate with 50 to 60% savings.

- **Mixed packages:** In mixed packages, the process flow includes automatically inserting through-hole components, overturning the board, and applying adhesive on the secondary side, and placing the SMT components. The adhesive is then cured and the board is overturned again. After the through-hole components are manually inserted, wave soldering is done on the secondary side. Cleaning, inspection, touch up, testing and rework follow. In mixed packages, a single-layer board is used for low-density, low-cost functions saving 25% of the real estate.

Soldering processes can now be completely automated. Solder paste is applied by dispensing machines. Components are placed with the help of chip shooters, and automatic placers incorporating computer vision systems identify where components should be placed. Reflow is a completely automated process, with sensors and timing devices to control the lengths of the various cycles. Wave soldering machines can now automatically flux, preheat, and wave-solder the components on the board, as well as clean the soldered board. Automatic stand-alone cleaning machines are also available. Inspection and rework are now both semi-automatic and fully automatic, and utilize machine vision. Even faster and more efficient systems are on the horizon. Figures 1.3 and 1.4 show various types of wave and reflow soldering machines currently available.

One of the greatest challenges of soldering technology is the development of reliable soldering processes for fine-pitch applications. Typical board-level fine-pitch applications involve lead pitches as close as 1 mm; chip-level soldering involves pitches as fine as 0.25 mm. Section 2.3.4 describes solders and solder pastes used for fine-pitch applications, while Section 4.6.3 covers fine-pitch soldering in more detail.

1.4 Soldering alternatives

Concern about environmental issues has prompted the soldering industry to search for alternatives to soldering as the primary interconnection technology

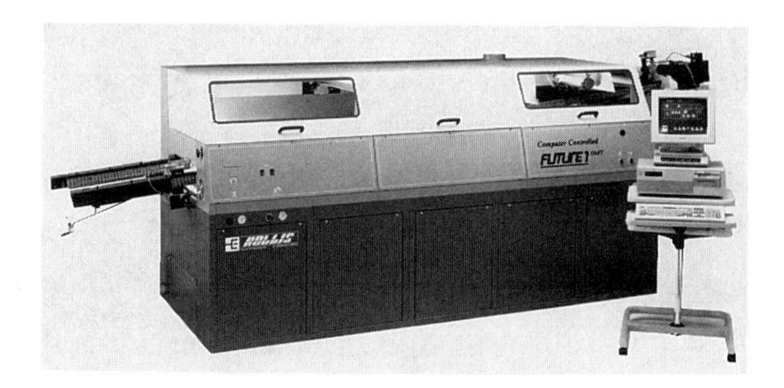

Figure 1.3: Wave soldering machinery (*Hollis Automation*)

for electronic assemblies. The principal environmental issues are raised by the use of chlorofluorocarbons (CFCs) for cleaning and of toxic materials, such as lead, in solder alloys, because of the role played by these materials in depleting the ozone layer and polluting the environment to an extent as to upset the natural balance of ecology.

One alternative to soldering is the use of conductive surface-mount adhesives (CSMAs), electrically conductive thermoplastic or thermoset resins filled with silver or gold particles 10 to 20 micrometers in diameter. Like solder paste, CSMAs can be printed using stencils and screens, and then be cured using heat and catalysts. However, CSMAs are harder to rework than solders and their adhesive characteristics on different metallic surfaces are not yet clear. Silver or gold depletion due to ionic migration in the presence of moisture and electric bias also poses a potential stumbling block to widespread use of CSMAs. However, CSMAs have certain specific advantages over solders, including the following:

- they facilitate fine-line resolutions for fine-pitch technologies;
- they have high flexibility, creep resistance, and stress dampening, and are therefore fatigue resistant;
- they require lower operational temperatures (120 – 150°C); and
- they pose no known environmental problems.

Another emerging solderless interconnection technology, fuzz buttons, consists of buttons populated in a molded engineering carrier (called vectra), with

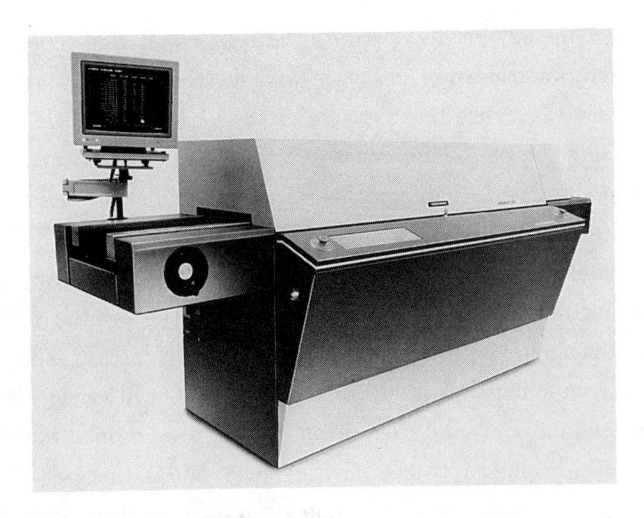

Figure 1.4: Reflow soldering machinery (*Electrovert*)

a hole pattern matching the electrical pad pattern on both the bottom of the chip carrier and the top of the printed wiring board. The typical button is a continuous gold-plated beryllium copper wire randomly woven into a cylindrical shape, as shown in Figure 1.5. The buttons are typically 0.020 in. in diameter and 0.040 in. in length; they protrude approximately 0.005 in. from both sides of the carrier. The electrical interconnection is formed by sandwiching the button carrier between the chip carrier and the printed wiring board.

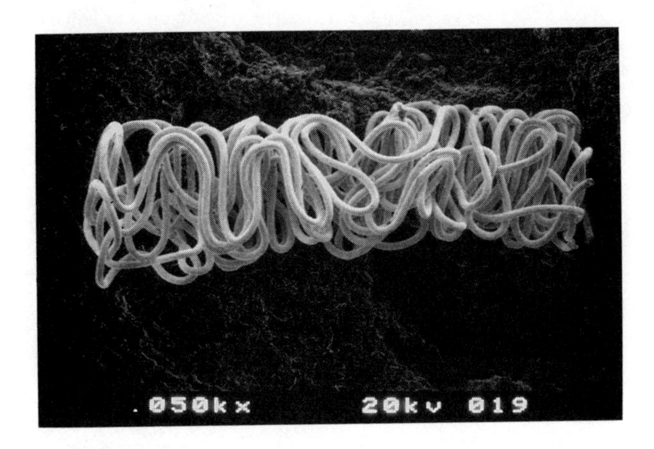

Figure 1.5: A SEM micrograph of a typical fuzz button

Button technology readily permits the use of pad-area-array component packages. Like pin-grid-array packages without the pins, the pad-area-array package can use the entire bottom area of the package for I/Os, thus significantly increasing the packaging density over comparable perimeter I/O packages. Solderless button technology also factors out solder process requirements such as high temperatures, solvents, and cleaning.

Westinghouse Electric Company has patented a technology that sandwiches fuzz buttons between the chip carrier and the printed circuit board. The fuzz buttons are fixed in a non-conducting button carrier, according to the layout of the components on the board. Component cavities are carved out in the module assembly unit and the button carriers are dropped into them. The components are dropped into the cavities next, and a thermal slug is tightened over the components to conduct heat away from them and provide the necessary compression for proper contact formation between the components and the pads below. The button carrier acts as a fixed stop, preventing further compression of the buttons.

Chapter 2

Solders, Solder Fluxes, and Solder Pastes

S.M. Bhandarkar
CALCE Electronic Packaging Research Center
University of Maryland
College Park, MD

2.1 Solders

Solders are alloys with melting temperatures (liquidus) below 400°C, formed from elements such as antimony (Sb), bismuth (Bi), cadmium (Cd), gold (Au), indium (In), lead (Pb), silver (Ag), and tin (Sn). (The melting points of some of the elements used in solder are shown in Table 2.1.) The commonly used alloys for electronic applications are binary combinations of tin and lead, with traces of other elements to tailor solder properties for specific applications, although solders based on other elements — such as bismuth, gold, and indium, among others — are also used. Table 2.2 shows commonly used solder compositions [Federal specification QQ-S-571] and associated solidus and liquidus temperatures. The solder alloy has different properties than those of its constituents.

Typically, solders are formed from a eutectic composition of alloy (see section 2.1.1). The principal advantages of using an eutectic alloy over other compositions are (1) immediate solidification at eutectic temperature decreases the chances of component movement or disturbance during cooling; (2) soldering can be achieved at the lowest possible temperature for the given material

	Sn	Pb	Ag	Bi	Cd	In	Sb
°C	231.9	327.5	960.5	271	321	156.4	630.5

Table 2.1: Melting points of metals used in solder

Alloy	Solidus (°C)	Liquidus (°C)
96.5Sn-3.5Ag	221	221
95Sn-5Sb	234	240
70Sn-30Pb	183	193
63Sn-37Pb	183	183
62Sn-36Pb-2Ag	179	179
60Sn-40Pb	183	191
50Sn-50Pb	183	216
42Sn-58Bi	138	138
40Sn-60Pb	183	238
35Sn-63Pb-2Sb	185	243
30Sn-68.5Pb-1.5Sb	185	250
20Sn-79Pb-1Sb	184	270
10Sn-88Pb-2Ag	268	290
10Sn-90Pb	268	301
5Sn-95Pb	308	312
1Sn-97.5Pb-1.5Ag	309	309
97.5Pb-2.5Ag	304	304
94.5Pb-5.5Ag	304	380

Table 2.2: Commonly used solder alloys (Federal specification QQ-S-571)

combination; and (3) non-eutectic compositions require higher temperatures to reach liquid solution.

Alloy selection is based on several factors, including electrical properties, alloy melting range, wetting characteristics, resistance to oxidation, mechanical and thermomechanical properties, formation of intermetallics, and ionic migration characteristics. These properties are important, because they determine whether the solder joint can meet the mechanical, thermal, chemical, and electrical demands placed on it. Solder alloys, fluxes, and their properties will be discussed in detail in this chapter.

2.1.1 Tin-lead alloys

The popularity of tin-lead alloys is primarily due to their melting range, good wetting characteristics, and affordable prices. The melting points of pure tin and lead are 232°C and 328°C, respectively. Depending on the composition, the alloy can exist as various phases at different temperatures: pure solid, solid solution, liquid solution, or an intermediate solid-liquid (pasty or plastic) phase, as shown by the phase diagram in Figure 2.1. The solidus and liquidus temperatures are those at which the alloy makes a transition from the pasty phase to the solid and liquid phases, respectively. Eutectic alloys go directly from solid to liquid, or vice versa, without going through the intermediate pasty phase. The solidus and liquidus temperatures are the same for the eutectic composition and represent the lowest melting temperature of the alloy for any composition.

Solder alloys with 63% tin (63Sn-37Pb) have been traditionally classified as eutectic solders. The correct eutectic tin-lead composition is 61.9% (by weight) of tin, with a eutectic temperature of 183°C. However, due to the presence of impurities, the eutectic composition is not exact. In practice, the proportions of solder alloys vary in the range of 63Sn-37Pb; 60-40 tin-lead alloys are commonly used in place of true eutectic solder, due to the lower price of lead. The use of non-eutectic alloy compositions can lead to separation of one of the phases from the solution upon solidification, causing dendrites and other particulates to form in the solder.

In general, lead takes little part in the reactions at the interface (wetting and intermetallic formation), but affects the solidus temperature of the alloy. Tin is more reactive and plays a major role in metallurgical reactions with metals. High lead content solders have higher solidus temperatures and are used

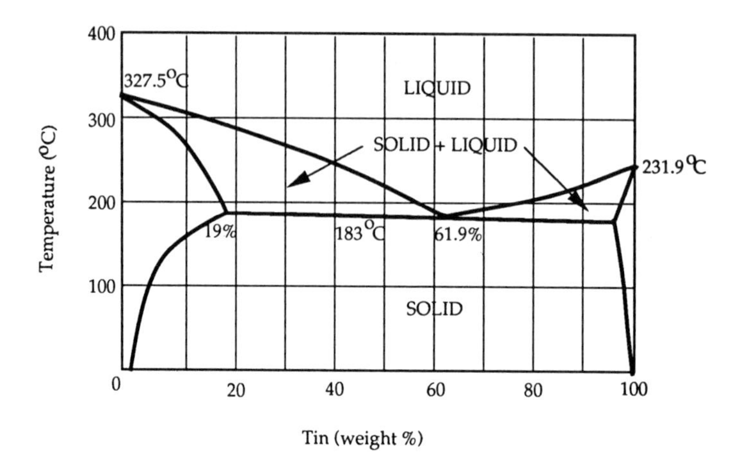

Figure 2.1: Tin-lead binary phase diagram

in high-temperature applications, such as step soldering to hold subassemblies in place, while lower solidus temperature solders are used for other joining operations.

2.1.2 Other alloys

Metals such as antimony, bismuth, cadmium, gold, indium, and silver, among others, are added to tin and/or lead to form binary or ternary alloys with properties suitable for specific applications. The goal is to develop alloys with improved creep resistance, tensile strength, fatigue strength, and other beneficial physical and metallurgical properties. The phase diagrams of several important binary alloys for soldering can be found in Hansen [1958].

Tin-lead-silver

A eutectic ternary composition, 62Sn-36Pb-2Ag, is used when soldering to silver-coated surfaces to decrease dissolution of surface silver into the solder. Complete silver dissolution from the surface will leave the surface bare, resulting in an unsolderable surface. The deliberate addition of silver to the solder melt prevents the dissolution of surface silver. Silver addition is also beneficial in reducing creep. However, some tin-silver intermetallics (e.g., Ag_3Sn) are detrimental to the strength of the solder interconnection. Typical applications of these solders are in printed circuit assemblies, component assemblies and thick- and thin-film silver hybrids.

Tin-lead-antimony

Antimony is added to tin-lead alloys to improve solder strength, but too much antimony can result in wetting deterioration and the formation of brittle antimony-tin intermetallics that are harmful to the solder joint. Tin-antimony binary alloys, such as 95Sn-5Sb, exhibit high creep resistance. These solders are typically used in component assemblies and palladium and silver thick- and thin-film hybrids.

Bismuth alloys

Alloys containing bismuth have low solidus temperatures (50° to 150°C) and are referred to collectively as fusible alloys. The ternary 60Bi-27Pb-13Sn and binary 42Sn-58Bi alloys are eutectic compositions with melting temperatures of 70°C and 138°C, respectively. They are used extensively to solder heat-sensitive devices. However, the wettability of these solders is poor, making proper cleaning or tinning necessary prior to soldering. Due to their low solidus temperatures, most fusible alloys are prone to creep.

Gold alloys

Gold-tin alloys exhibit high strength and excellent corrosion resistance and are typically used in die attach and hermetic seals. The commonly used gold-tin alloys are 80Au-20Sn eutectic and 96.7Au-3.3Sn. However, gold solders are very expensive and require special equipment with inert atmospheres to produce acceptable solder joints. Inert atmospheres are required to prevent oxidation of workpieces at the high solidus temperatures of these solders (280°C and above).

Indium alloys

Indium solders provide good solderability, ductility, and thermal fatigue properties. Tin-indium alloys are used for soldering glasses and ceramics. Lead-indium alloys have lower rates of gold dissolution than tin-lead alloys and are used in gold thick- and thin-film hybrids. Indium solders have a wide range of solidus temperatures (44In-42Sn-14Cd and 5In-92.5Pb-2.5Ag melt at 93°C and 310°C, respectively), and are widely used for step-soldering operations.

2.1.3 Intermetallics

Solders are used to join a variety of materials, including copper, aluminum, nickel, gold, and silver. The materials that compose the workpieces to be soldered are referred to as base elements. Intermetallic compounds are generally formed between the base elements and the solder by liquid-solid reactions and/or solid-state diffusion, either during the soldering operation or in service. The formation of intermetallics at the interface of the materials being soldered is the basis of the solder bond. However, intermetallic formation can proceed into the body of the solder joint as well, altering its mechanical properties. Intermetallics are typically formed through ionic bonds, which make them hard and brittle in comparison with the parent solder alloy. Although most intermetallics are tough, Ag_3Sn and $AuSn_x$ have low strength and should be avoided in electronic solder attachments.

Intermetallics have a definite stoichiometric composition, as shown in Table 2.3. Aluminum and zinc do not form intermetallics with tin or lead and have been omitted from the table. Cadmium forms an intermediate phase that decomposes below 130°C and has also been omitted. Tin is more reactive than lead and forms more intermetallics with the base elements. Since, copper and gold are common base metals used in electronics, tin-gold and tin-copper intermetallics have been widely studied [Frear 1991].

The amount of intermetallics formed at the interface during soldering depends on the solubility of the solder elements in the base metal(s), and on the time and temperature of the soldering operation. Intermetallics are formed in the body of the solder as the solder solidifies and as the base metal(s) dissolved in the molten solder precipitates out. Subsequent growth of intermetallics during storage and operational service occurs through a solid-state diffusion process described, generally, by an empirical equation [Frear 1991]:

$$x(t,T) = x_\circ + A\,t^n e^{-Q/RT} \tag{2.1}$$

where x is the total intermetallic thickness at time t and temperature T (Kelvin), x_o is the thickness of the intermetallic at time $t = 0$ (at the end of the soldering

	Sb	Bi	Cu	Au	In	Fe	Ni	Pt	Ag
Pb		$BiPb_3$		Au_2Pb $AuPb_2$	In_3Pb			Pt_3Pb $PtPb$ $PtPb_4$	
Sn	SbSn		Cu_3Sn Cu_6Sn_5	AuSn $AuSn_2$ $AuSn_4$		FeSn $FeSn_2$	Ni_3Sn Ni_3Sn_2 Ni_3Sn_4	Pt_3Sn PtSn Pt_2Sn_3 $PtSn_2$ $PtSn_4$	Ag_3Sn

Table 2.3: Intermetallic compounds formed between tin-lead and common solder elements and impurities [Manko 1979]

operation), R is the universal gas constant, Q is the activation energy, and A and n are constants. The linear growth rate, $n = 1$, describes many highly reactive systems (gold in contact with tin-lead-indium alloys and copper in contact with tin-indium alloys [Frear 1991]). Parabolic growth rates, $n = 0.5$, can be used to approximate intermetallic growth controlled by bulk diffusion and declining with time (copper in contact with many tin-lead alloys) [Frear 1991]. Some systems display sub-parabolic growth rates, such as $n = 0.37$ for copper in contact with 60Sn-40Pb alloys.

2.1.4 Contaminants in solder alloys

Impurities in solder result either from contaminants in the raw materials or from dissolution of metals (from assemblies and equipment) into the solder during the soldering operation. The equipment used in soldering is typically constructed from metals that are insoluble in molten solder and not wetted by it. Therefore, impurities picked up during the soldering operation come mostly from the assemblies being soldered (copper, nickel, silver, and gold from components and boards). Table 2.4 shows the maximum acceptable contamination limit in solder baths for some common impurities [IPC-S-815]. The limits in Table 2.4 were established from observations in manufacturing environments, since a comprehensive study for all possible contaminants is a difficult undertaking. However, the effect of single impurities on tin-lead solder have been studied under controlled conditions by several researchers [Ackroyd et al. 1975, Becker and Allen 1970]. The results are summarized in the following paragraphs.

Contaminant	Contaminant Limit %		Characteristic effect of contaminant on solder
	Preconditioning (lead/wire tinning)	Assembly (Pot, Wave, etc.)	
Cu	0.750	0.300	Sluggish and gritty solder with increased bridging
Au	0.500	0.200	Sluggish and gritty solder
Cd	0.010	0.005	Sluggish, porous solder with oxidation on surface
Zn	0.008	0.005	Rough, grainy, frosty, porous solder high dendritic structure, surface oxide
Al	0.008	0.006	Sluggish, frosty, porous solder surface oxide on pot
Sb	0.500	0.500	Sluggish solder
Pb	0.020	0.020	Gritty solder, resoldering problems
As	0.030	0.030	Sluggish solder, blisterlike spots
Bi	0.250	0.250	Lowered melting point, surface oxide
Ag	0.750	0.100	Possible reduced fluidity
Ni	0.025	0.010	Brittle solder, hard insoluble compounds

Table 2.4: Contaminant limits for solder baths [IPC-S-815]

Aluminum

The source of impurity is typically the aluminum fixtures used during soldering. Aluminum has very little solubility in solder, and only small amounts (0.5%) can be dissolved in solder at soldering temperatures. The presence of aluminum causes sluggishness in the solder melt and a gritty, dull appearance of the solder joint due to aluminum oxidation. The maximum acceptable impurity level of aluminum in the melt is set at 0.005%. Ackroyd et al. [1975] have reported that the addition of small amounts of antimony to the solder results in the formation of an aluminum- antimony intermetallic that is removed from the melt with the dross (oxide scum and other wastes).

Antimony

Antimony is intentionally added to solder alloys to prevent transformation of tin at very low temperatures into a powdery grey state, called tin pest. The presence of antimony also offsets the ill effects of aluminum, as explained above. ASTM specifications set the maximum antimony content at 0.5%. Small additions of antimony, up to 0.3%, have been shown to improve the wetting properties of solder, but larger additions slowly degrade wetting. Apart from intentional addition, other sources of contamination are unlikely.

Arsenic

Raw material is the only source of arsenic contamination. Arsenic contamination is not considered a major problem, since it is present in very small quantities (0.05% or less). Arsenic impurities cause dewetting of brass surfaces. Arsenic's intermetallics with tin-lead form long needles that may degrade mechanical properties.

Bismuth

Bismuth is added to tin-lead solders to improve wetting characteristics. The amount added is limited to 0.25% in raw tin-lead solder material. Bismuth base alloys with up to 67% bismuth content form a wide class of low-temperature fusible solder alloys. Alloys containing more than 47% bismuth expand upon cooling [Manko 1979]. This anomalous expansion can be used to reduce stresses caused by thermal expansion mismatches in electronic assemblies.

Cadmium

The presence of cadmium in the solder melt may promote the formation of oxides and tarnishes and make the melt sluggish. A limit of 0.001% is imposed on cadmium impurities in tin-lead solders. Like bismuth, cadmium-based alloys are used as low-temperature fusible solders.

Copper

Copper is the most common contaminant in solder melts because most leads, traces, and vias are copper plated, and contamination of the bath during soldering is difficult to avoid. Copper forms hexagonal needle-shaped intermetallics (Cu_3Sn and Cu_6Sn_5) with tin at room temperatures. The intermetallics cause sluggishness in the solder melt, grittiness in the solder joint, and decreased wetting ability. Therefore, copper must be limited to 0.3% in the melt, after which the bath is treated or replaced.

Gold

Gold contamination arises from gold platings on the workpiece. Gold has a high dissolution rate in solder, forming several intermetallics with tin and lead (Au_2Pb, $AuPb_2$, Au_6Sn, $AuSn$, $AuSn_2$, and $AuSn_4$). These intermetallics are brittle and can lead to solder cracking. Gold contamination levels of above 0.5 percent cause sluggishness of the melt and reduce wetting ability.

Iron

Iron forms intermetallics with tin ($FeSn$ and $FeSn_2$). The primary source of contamination is the solder pot, which is often made of cast iron. However, at typical soldering temperatures, the rate of dissolution of iron in solder is low. The presence of iron causes grittiness in solder joints.

Nickel

Nickel has very little solubility in tin-lead, and the only source of contamination is the nickel coating on some component leads. Nickel causes gritty solder joints.

Phosphorus

Phosphorus impurities result in tin and lead phosphides, causing dewetting and gritty joints. Contamination in the bath may result from electrodeless nickel platings that contain some phosphorus.

Silver

Silver is added to solder to form special solder alloys. When present in large amounts, silver may cause grittiness and formation of pimples on the surface due to the intermetallics Ag_6Sn and Ag_3Sn. The source of contamination in the bath is usually the workpiece being soldered.

Sulfur

Sulfur contamination results in dewetting, due to the formation of tin and lead sulfides. The source of sulfur is raw solder materials.

Zinc

When present in solder joints, zinc oxidizes rapidly on the surface, reducing the repairability of the joint (because zinc oxides do not react with fluxes). When brass is soldered, copper and nickel overplates are used to prevent migration of zinc to the surface. Zinc also forms excessive dross in the melt, leading to metal loss.

2.1.5 Solder forms

Solder materials are available commercially in many forms. The selection of a particular form is governed by the method used for soldering.

Solder pastes (creams)

Solder pastes combine solder alloy powder, flux, and binder in paste form. The paste is deposited at the soldering site and then reflowed to form the joint. Solder pastes also serve as an adhesive to hold surface-mount components in place prior to the soldering operation.

Solder pigs, ingots, and bars

Solder pigs are rectangular blocks of solder of varying sizes. Pigs typically weigh from 20 to 100 lb., while ingots and bars are smaller, weighing 2 to 20 lb. and 0.5 to 2 lb., respectively. These forms are used to feed solder into solder baths.

Solder wire, flux-core wire

Solder wires are continuous lengths of wire of various diameters available in spool or bulk form. Flux-core wires have a central core of flux. Wires are typically used in hand-soldering operations.

Solder preforms

Preforms are specially shaped solid forms (e.g., rings, cylinders, and spheres) customized for a specific joint geometry. The preforms are positioned at the soldering site and melted during the soldering operation to form the joint.

Solder sheets, foils, and ribbons

Solder is also available in sheets, foils, and ribbons of various sizes and thicknesses. These are available with or without flux coatings.

2.2 Fluxes

For the formation of a good solder bond between two metals, the metallic surfaces must be clean and free of oxides and tarnish. Because it is difficult to keep surfaces from oxidizing from the time of manufacture to the time of soldering, fluxes are used to chemically clean surfaces during the soldering operation. In general, fluxes do not remove dirt, oils, greases, or fingerprints.

Flux cleans oxides and tarnish from surfaces, increases wetting of surfaces by decreasing surface tension, and prevents re-oxidation of surfaces during the soldering process. In addition to these essential properties, the flux must flow out of the way of molten solder to enable formation of the solder joint, and must be stable enough to withstand high soldering temperatures without chemically breaking down. In addition, the residue left behind after soldering must be non-corrosive and non-conductive, or else be easy to wash away.

The selection of fluxes depends on several factors: type of metallic surfaces to be soldered, cleanliness of the surface (a more reactive flux is required for a dirty surface), and corrosivity of the residue. Solubility of the residue in water is also desirable if the residue is to be washed away after the soldering process.

Often, specifications govern the type of flux used. A highly reactive flux is desirable under poor solderability conditions to ensure proper soldering. However, these fluxes tend to be more corrosive and are not used in applications where the residue is difficult to wash away completely. For example, flux residues can get trapped in crevices, under components, inside connectors, and in unsealed components, such as switches and relays.

Some important properties of fluxes are listed below.

- **Flux activity and corrosivity.** Flux activity is the ability of the flux to wet and clean surfaces chemically. Corrosivity is the detrimental chemical attack of the flux and its residue on the surfaces being soldered. The higher the flux activity, the more corrosive the fluxes and their residues. Flux activity and corrosivity is measured by tests on the flux prior to the soldering operation (halide content, acid number, water extract resistivity, and copper mirror tests) or by tests on the board after the soldering operation (surface insulation resistance test). Details of these tests are contained in federal specification QQ-S-571. Flux activity increases with temperature until a temperature is reached at which the flux breaks down chemically. A flux with high activity is required for poorly solderable surfaces to achieve adequate wetting of the surface. For reliability over an extended period of time, any residues left behind after soldering must be non-corrosive or easy to wash away.

- **Cleanability.** Cleanability of the flux relates to the ease with which residues can be removed after soldering and is determined by solubility characteristics in water and solvents. Rosin-based fluxes are soluble in chlorofluorocarbon solvents. These fluxes can be made soluble in water by saponification (conversion into soap). The tackiness of residues is also an important factor in determining flux cleanability. For applications requiring cleaning, soft residues are desirable; for no-clean applications, hard, non-tacky residues, which encapsulate the contaminants, are desirable.

Other important properties of fluxes include viscosity, rheology, volatility, and thermal stability. These properties are strongly affected by additives to the flux and are discussed in Section 2.3.

2.2.1 Flux composition

Fluxes are composed of a base (or vehicle) and an activator. The vehicle can be solid or liquid, and acts principally as the carrier for the active constituents of the flux, such as activators, solvents, detergents, and other additives. In addition, the vehicle functions as a solvent to carry away the residues of fluxing and forms a protective layer over the cleaned surfaces to prevent re-oxidation by air. The vehicle itself may provide mild fluxing action, as in the case of rosins. Typically, however, the fluxing action is performed by the activators present in the flux.

An important category of vehicles are rosins and resins, which are composed of either naturally available compounds, such as rosins (from pine tree sap), or synthetically manufactured derivatives of rosins and resins. Another class of vehicles is water-soluble materials, such as alcohols, glycols and glycerols, which form the base for water-soluble fluxes that enable the residue to be washed away by water. Activators ranging from mildly active materials, such as organic salts, to highly corrosive chemicals, such as inorganic acids, provide chemical fluxing action to remove oxides and tarnish. Examples of mild activators are fatty acids and organics containing carboxylic and amine groups. The more aggressive fluxes contain halide salts of amines, such as amine hydrochloride. The most corrosive activators are organic acids, such as hydrochloric, hydrobromic, and hydrophosphoric acids. The type of activators used is determined by the surfaces to be soldered and the type of contamination to be removed. For example, halides are effective against copper oxides, organic acids reduce tin and lead oxides, and amines are best on silver surfaces.

Additional materials are added to the flux depending on the manufacturing process. Solvents like alcohols and glycols are used in wave soldering to carry flux to the surface of the assembly being soldered. These solvents are volatile and evaporate during preheating. Non-volatile solvents are used in solder pastes to control the rheology of paste, an important property both during solder reflow and during application of solder paste onto solder pads. Surfactants are added to some fluxing machines to generate foam in the flux.

2.2.2 Types of fluxes

The composition or type of vehicle, level of activity of the residue, and cleaning requirements are all important in defining fluxes. A widely accepted classifica-

Test	RMA flux	RA flux
Water extract resistivity	100,000 Ω.cm minimum	50,000 Ω.cm minimum
Silver chromate test (halides)	No chloride, bromide response	Not required
Copper mirror test	Incomplete removal of copper film	Not required

Table 2.5: MIL-F-14256 specification for RMA and RA fluxes

tion for fluxes involves three broad categories: rosin-based fluxes, water-soluble fluxes, and synthetically activated fluxes. Specifications for rosin-based fluxes are covered by MIL-F-14256.

Another classification of fluxes is based on the conductivity and corrosivity of fluxes and their residues [IPC-SF-818]. The fluxes are classified as L (low), M (moderate) and H (high), where L, M, and H denote the level of corrosivity and electrical conductivity. The classification by the IPC applies to all fluxes and is not limited to rosin-based fluxes alone, as in the case of the MIL-F-14256 specification.

Rosin and rosin-based fluxes

Fluxes which use rosin as the vehicle are the least active. Rosins are naturally occurring substances obtained from gum or sap extracted from pine trees. When rosin is used by itself (R-flux or white-water rosin), fluxing action is provided by mild abietic and primaric acids present in the rosins. The weak action of the flux necessitates cleaning prior to soldering. Activators are added to improve the fluxing action. Rosin mildly activated (RMA) fluxes have activators that leave behind non-corrosive and electrically non-conductive residues (low ionic content). The activators in this category include organic acids, halides and amides. Fully activated rosin (RA) fluxes are more active than RMA fluxes and are permitted higher ionic content in their residues. The typical specifications for corrosivity and ionic content for types RMA and RA fluxes are shown in Table 2.5 MIL-F-14256.

Rosins are sometimes modified by chemical processes, and the end products are used in fluxes. Some manufacturers refer to these as resin fluxes. This term is also loosely applied to rosins that are not classified by military standards and to some synthetic resin products.

Rosin-based fluxes are widely used because of their favorable properties. Besides having a mild fluxing action, the rosins form a hard barrier largely impermeable to ions and moisture. The mild acids present in the rosins are inactive at room temperature and do not present corrosivity problems, thus enabling low-activity rosin fluxes to be used in surface-mount applications, where removal of residues between the board and the component is difficult.

Rosins are soluble in a wide variety of organic non-polar solvents that facilitate application of flux and subsequent cleaning. Rosins can be made water-soluble by using saponification agents to convert the rosins into soap to facilitate cleaning.

Water-soluble fluxes

Water soluble fluxes, also referred to as organic intermediate fluxes, organic water washable fluxes, and organic acid fluxes, are composed of materials whose residues can be washed away by water. Water-soluble fluxes are both ecologically safe and economically viable. In general, water-soluble fluxes contain a higher level of organic acids and halides than the most active rosin flux. Due to the presence of these easily ionizable compounds, residues must be washed away, or they may result in corrosion and leakage currents in humid and high-temperature environments. The design of the assembly (e.g., the use of sealed components, and proper spacing between components and between components and the board) must ensure that the flux does not get trapped in the assembly.

Synthetic-activated (SA) fluxes

Synthetically-activated fluxes are soluble in non-polar organic chlorofluorocarbon solvents but provide the high fluxing activity of water-soluble fluxes, thus enabling the soldering of hard-to-solder surfaces using traditional chlorofluorocarbon solvents. As with water-soluble fluxes, the residues are corrosive and must be washed away.

Other fluxes

Fluxes based on inorganic acids, alkalies, and their salts are extremely corrosive. Hydrochloric acid, orthophosphoric acid, zinc chloride, ammonium chloride, and other halides are examples of inorganic fluxes. Due to their high corrosivity, these fluxes are not used for electronic soldering, except for hard-to-solder materials, such as leads made of kovar or nickel-plated copper. All residues must be removed after soldering.

To reduce the cost associated with cleaning, a class of 'no-clean' fluxes that leave little or no residue after soldering has been developed. These fluxes ensure cleanliness under surface-mount components and other hard-to-inspect spots. The no-clean fluxes, mildly activated fluxes with a low solids content (about 15%), leave little or no residue after soldering.

Soldering is sometimes performed in controlled atmospheres, either reactive or protective. Reactive atmospheres containing oxidizing and/or reducing agents add to the fluxing action. Protective or inert atmospheres protect the materials being soldered from oxygen, moisture, and other atmospheric constituents that may oxidize or contaminate the solder joint.

2.3 Solder paste

Solder paste (or solder cream) plays an important role in automated soldering of surface-mount components. Solder pastes are composed of a mixture of solder alloy powder, flux, and other additives. They provide the desired rheology and adhesive characteristics for surface-mount applications. When in the viscoelastic form, solder paste can be automatically applied at selected sites in different sizes and shapes. The adhesive characteristics permit the attachment of surface-mount components to the board prior to solder reflow.

2.3.1 Solder alloy powder

Solder alloy powder is manufactured from bulk solder alloy by several techniques: chemical reduction, electrolytic deposition, mechanical processing of solid metal, and atomization of molten metal. The powders used in soldering are typically produced by atomization, which involves injecting liquid metal under high pressure through small orifices into gaseous or vacuum chambers. The process produces powders with spherical particles that are desirable for solder pastes. The size distribution of the powder is controlled by the feed rate, orifice design, molten metal temperature, and gas pressure in the atomization chamber.

An alternative method of powder manufacture involves the use of a spinning disc. Molten metal is fed to the disc and spun off the knife edges. Although temperature and size of the chamber, velocity of the disc, and feed rate are adjusted to produce spherical particles, a large number of elongated particles tend to be produced.

Chemical reduction manufacture of solder powder involves precipitation of metals from aqueous solutions of their salts. The resulting powders are spongy and porous. Electrolytic deposition is another solder powder manufacturing method that produces high-purity powders.

Some of the characteristics of solder alloy powders are particle shape, particle size distribution, flow rate, apparent density, and oxide content. These characteristics are discussed further below.

Particle shape

Spherical particles with smooth surfaces, as produced by atomization, are desirable for solder pastes. Large numbers of irregularly shaped elongated particles interfere with printing operations. Sometimes satellites of smaller particles become attached to larger particles, forming clusters, and result in unwetted surfaces and creation of solder balls.

Particle size and distribution

Standard wire meshes [ASTM B-214] are used to determine particle sizes of powders by sieve analysis. Typically, two mesh sizes, -200/+325 and -325, are used for solder pastes. These correspond to particle sizes in the range of 45/75 microns and less than 45 microns, respectively. The shape of the particles must be determined optically before sieve analysis, because the presence of a large number of irregularly shaped particles may skew the results of the analysis. The weight distribution of particles of different diameters can be determined by several techniques, including light scattering, sedigraph, air-jet classification, and electrical zone sensing [Hwang 1989]. Desirable particle size distributions for solder pastes [IPC-SP-819] are given in Table 2.6.

Solder paste type	Nominal particle size in μm (percentage by weight)		
	< 1% must be larger than	at least 80% must be in the range	at most 10% must be less than
Type 1	150	75 - 150	20
Type 2	75	45 - 75	20
Type 3	45	20 - 45	20
Type 4	38	20 - 38	20

Table 2.6: Solder-powder particle size distribution for solder pastes [IPC-SP-819]

Flow rate

The flow rate is determined, according to ASTM B-213, by pouring a given weight of powder through a Hall flowmeter and measuring the time taken for the powder to run through. If the powder is too fine or contaminated, the particles stick together, hindering the flow rate.

Apparent density or tap density

The density of powder is lower than the density of fused alloy, due to the packing of the particles in the powder. The apparent density of powders produced by atomization (measured by ASTM B-212) is approximately half the true density of the bulk alloy. The apparent density of powder determines the amount of powder required to create a solder joint of a given volume.

Oxide content

Powder is manufactured in environments minimizing oxide formation — for example, gas atomization, in which a nitrogen atmosphere is used. However, during subsequent handling, storage, and processing operations, the powder must remain oxide-free. The oxide content and contamination levels are determined by fusion tests and oxygen detectors.

2.3.2 Solder-paste flux

The same traditional flux formulations described in Section 2.2 are used as solder-paste fluxes, with modifications for the paste application process. For example, solder pastes typically contain non-volatile solvents, that do not dry out during printing, as opposed to traditional fluxes with volatile solvents. Very low-activity fluxes using rosins and resins as the only activators are generally not used in solder pastes. The most widely used fluxes are of the RMA and RA type, with a variety of supplementary activators. Specifications such as QQ-S-571 and MIL-F-14256 require complete removal of RA flux residues.

Solder pastes are typically used with surface-mount components, increasing the risk of trapping corrosive flux residues under components. Therefore, RMA fluxes with milder activators are a better alternative than RA fluxes. Solder-paste fluxes are formulated to leave either soft or hard residues. Soft residues are easy to clean, while hard residues form an encapsulant for the ionic contaminants and need not be cleaned.

Fluxes used in solder pastes contain additives such as solvents and thickeners to modify the viscosity and rheology (flow and deformation properties) of the paste. The solvents dissolve the solids in the flux and, together with the thickeners, maintain the correct rheology during paste dispensing and application, ensuring the desired shape and volume of solder paste. The rheology must be suited to the reflow process to avoid solder slump and voids. Other additives to solder paste include tack modifiers and wetting agents. Tack modifiers ensure that paste tackiness is retained for the period between paste application and component placement.

2.3.3 Paste

Solder paste is a complex homogeneous blend of alloy powder, flux and additives. The selection of the right blend depends on the particular application and requires a thorough understanding of the performance parameters and properties of the solder paste. Some of these parameters and properties pertaining to alloy powders and fluxes including particle shape and size, flow rate, apparent density and oxide content of alloy powders and activity, corrosivity, and cleanability of fluxes have already been discussed. Other characteristics — metal content, rheology, viscosity, solder slump, tackiness, solvent volatility — and thermal stability are summarized below.

Transfer method	Viscosity (Pa. s)	Metal content (%)
Syringe Dispensing	350 - 600	80 - 85
Screen printing	450 - 675	88 - 90
Stencil printing	550 - 750	90

Table 2.7: Typical paste viscosity and metal content for common paste-transfer methods [Capillo 1990]

Metal content

The metal content of solder paste is the percentage metal by weight in the solder paste. The volume percent of metal, V_m, calculated from the densities of the alloy and flux systems, determines the amount of solder paste required for producing a joint of a given height and shape. That is,

$$V_m = \frac{W_m}{W_m + \{\rho_m/\rho_f\}W_f} \tag{2.2}$$

where V is the volume, W is the weight percent, and the subscripts m and f stand for metal and flux, respectively. Increased metal content increases the viscosity of the paste, minimizing bridging and voiding in the solder joint. A high metal content is desirable, provided the increased viscosity does not hinder paste application. Typical metal contents in 60Sn-40Pb solders are 85 to 92% by weight. Table 2.7 shows typical metal contents for paste applications.

Rheology and viscosity

During application of solder paste and reflow, the rheology of solder paste governs deposition of the paste as well as the geometry of the solidified solder joint. Solder paste is a viscoelastic material, and therefore its characterization is rate-dependent. The viscosity of solder pastes decreases when subjected to increasing shear rate and increases again over time after the shear is removed, a time-dependent flow phenomenon called thixotropy. Rotational viscometry is used for measuring the shear stress and viscosity of solder pastes as functions of shear rate and time. Dynamic mechanical analyzers are used for measuring mechanical properties, such as storage and loss moduli as functions of frequency, strain amplitude, and time. Most rheological properties are also functions of temperature.

Due to the complexities involved in rheological measurement, federal specification QQ-S-517 mandates single-point viscosity measurement, using the Brookfield viscometer as the standard for characterizing solder paste rheology. Typical single-point viscosity values for some common paste-transfer techniques are given in Table 2.7.

Solder slump

Solder slump or spread is a flow characteristic of solder paste that affects the reflow process. As the temperature increases during reflow, the solder paste tends to slump or spread out, aided by component weight. Excessive slump can cause bridging between pads (a critical issue in fine-pitch technology) and less-than-optimal solder joint heights. The total slump of deposited solder paste during reflow is a combination of both solder paste slump and the flow of molten solder.

Slump is measured by cold-slump, hot-slump, and molten-flow tests. The first two tests relate to paste slump at ambient and elevated temperature; the third test takes into account only the flow of molten solder. Federal specification QQ-S-571 outlines requirements for the spread factor for solder pastes. Increased slump resistance can be achieved by using non-metallic spheres as spacers in the paste, by increasing the metal content, by including high-melting-point metal inclusions in the paste, and by using rheological additives in the solvent.

Tackiness

Pastes have inherent tackiness properties to enable surface-mount components to be held in position prior to solder-joint formation. The tackiness generally decreases with exposure to atmosphere, so the time between paste application and reflow must be held to a minimum. The tackiness of paste can be measured using the IPC-SP-819 test procedure, and can be improved with proper storage and handling.

Solvent volatility

Pastes may have either volatile or non-volatile solvents that determine if the paste can be dried prior to the solder melt. Pastes with volatile constituents can be dried by preheating before reflow, and thus have the advantage of reducing void formation and paste spattering during reflow.

Thermal stability

Because most constituents of the flux and activator system are organic, chemical breakdown can occur at elevated temperatures. Thermal stability is particularly important for high-lead-content solder requiring high reflow temperatures. If organics get charred, the cleaning of residues becomes more complicated and properties of the solder joint deteriorate.

2.3.4 Solder paste requirements

Traditional pitches on device pins range from 25 mils to 50 mils for surface-mount components. For fine-pitch devices, the pitch varies from 25 mils to as low as 4 mils. Fine spacing between solder pads requires refinements in paste application processes and modifications in solder paste characteristics to prevent bridging between pads during the soldering operation. In particular, the proper printing or dispensing methods and pastes with proper rheology and slump properties must be used. Desirable paste properties and processing parameters for fine-pitch technology are summarized below.

- The paste and molten solder must have high flow restrictivity at reflow temperatures to ensure that the paste and the molten solder do not bridge the gap between pads.
- The paste must be free from solder balls.
- The alloy powder must be of small uniform size (5 to 35 microns). The presence of larger particles mixed with the small particles results in uneven packing and causes slumping of the large, unsupported particles. Slumping causes bridging between lands. The smaller particles expose more surface area to the atmosphere, increasing susceptibility to oxidation.
- Flux residues must not restrict flow and must be easily cleanable.

Property	60Sn-40Pb	50Sn-50Pb	10Sn-90Pb	96.5Sn-3.5Ag	95Sn-5Sb	62Sn-36Pb-2Ag
Surface tension (mN/m)	481 @ 260°C	476 @ 280°C	437 @500°C	480	-	376
Viscosity (Ns/m)	0.01682 @350°C	0.01775 @ 350°C	0.02271 @350°C	-	-	-
Density (g/cm³)	8.52	8.89	10.5	7.29	7.3	8.4

Table 2.8: Physical properties of solder alloys [ITRI Publication No. 656, 1987; Hwang 1989]

- The volume of paste deposited on the pad must be tightly controlled. Too much paste causes bridging and too little results in solder joints with less than optimum height and cross-section.

- The component placement force must be tightly controlled. Excessive force may cause solder to flow out from under the component, resulting in bridging; insufficient force may result in improper soldering.

2.4 Properties of solders

The physical and thermal properties of solder play an integral role in the electrical function of the solder joint. In addition, the mechanical and thermomechanical properties of the solder joint are key factors in determining interconnection reliability. Published values of solder properties often vary from source to source because of differences in specimen design and casting conditions for the alloys.

2.4.1 Physical properties

The important physical properties of solder alloys are solidus and liquidus temperatures, surface tension, viscosity, and density. These are briefly discussed; values are provided for some common solder alloys in Table 2.8.

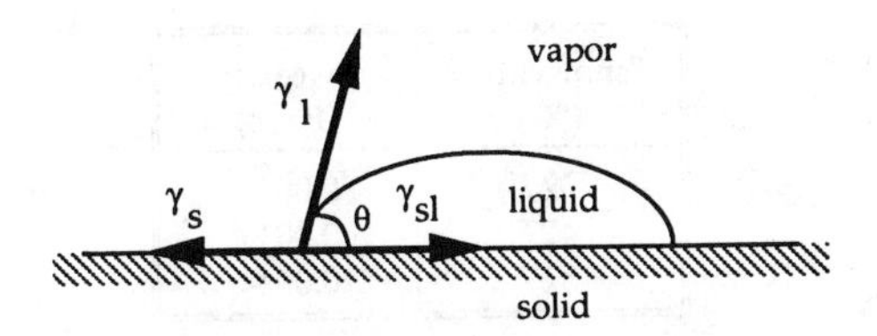

Figure 2.2: Surface tensions at liquid, solid and vapor interfaces

Solidus and liquidus temperatures

The solidus temperature is the temperature at which the solid alloy begins to melt (goes from solid solution to pasty or plastic phase, or vice versa). The liquidus temperature is the temperature at which the alloy melts completely (goes from pasty phase to liquid solution, or vice versa). The solidus and liquidus temperatures of common solder alloys are given in Table 2.2. For a given alloy, there exists eutectic composition at which the solidus and liquidus temperatures are equal; this represents the lowest melting point for the alloy.

Surface tension and wetting

The ability of the molten solder to wet surfaces is determined by the surface tensions between the various contact surfaces. Figure 2.2 shows the surface tensions acting at the interfaces of a liquid droplet on a flat solid surface. The angle, θ, formed between the liquid and solid surfaces is called the contact or dihedral angle. The surface tensions between solid-vapor, liquid-vapor, and solid-liquid, respectively, are represented as γ_s, γ_l, and γ_{sl}, which are inherent material properties. The condition for spreading (neglecting gravity) can be obtained from thermodynamics and is given by:

$$\gamma_s > \gamma_{sl} + \gamma_l \cos\theta \tag{2.3}$$

For good spreading or wetting during soldering, the surface energy of the metal to be soldered must be kept greater than that of molten solder. There is total wetting of the surface when $\theta = 0°$ and total non-wetting when $\theta = 180°$. For typical chip-to-board interconnections, $\theta = 75°$ is considered a good wetting angle. In general, for tin-lead alloys, the surface tension decreases with

Temperature (°C)	Viscosity (Pa. s)
327	0.0021
427	0.0017
527	0.0014

Table 2.9: Viscosity of 60Sn-40Pb alloys as a function of temperature [CINDAS report 102 1991]

increasing lead content. The values of surface tension in Table 2.8 are for liquid solder, that is, γ_l (liquid-vapor surface tension).

Viscosity

Viscosity refers to the property of molten solder that governs its flow during the soldering operation. For tin-lead alloys, viscosity increases with increasing lead content. The viscosity of an alloy can be computed from its density and molecular weight [Andrade 1952]. The viscosity for 60Sn-40Pb alloys decreases with an increase in temperature, as shown in Table 2.9 [CINDAS Report 102, 1991].

2.4.2 Electrical properties

Electrical conductivity

The electrical conductivity of solder alloys is typically ten times lower than that of copper and is measured as a percentage of the conductivity of copper. Electrical conductivity generally decreases with increasing lead content (see Table 2.10).

2.4.3 Thermal properties

Thermal conductivity

The thermal conductivity of solder is important when the solder joint is the primary path for conducting heat away from the package. The thermal conductivity of solder decreases with increasing lead content and is shown in Table 2.11 for common solder alloys.

Electrical conductivity (% copper IACS)	60Sn-40Pb	50Sn-50Pb	10Sn-90Pb	96.5Sn-3.5Ag	95Sn-5Sb	62Sn-36Pb-2Ag
	11.5%	10.9%	8.2%	14%	11.9%	11.6%

Table 2.10: Electrical conductivities of solder alloys [ITRI Publication No. 656, 1987; Hwang 1989]

Property	60Sn-40Pb	50Sn-50Pb	10Sn-90Pb	96.5Sn-3.5Ag	95Sn-5Sb	62Sn-36Pb-2Ag
Thermal conductivity (W/m. K)	50	46.5	36	33	28	-
Specific heat capacity (J/Kg/K)	150	210	-	-	-	-
Latent heat of fusion (KJ/Kg)	37	53	-	-	-	-
Coefficient of thermal expansion (ppm/°C)	23.9	23.4	27.98	28.7	29.6	23.6

Table 2.11: Physical properties of solder alloys [ITRI Publication No. 656, 1987; Hwang 1989]

Specific heat and latent heat of fusion

The specific heat is the heat required to raise the temperature per unit weight of the solder by 1°C. The latent heat of fusion is the amount of heat required to melt a unit weight of solder at constant temperature. Both quantities are used in determining the correct temperature and heating of the solder melt in machine soldering operations. The specific heat and latent heat increase with increasing lead content.

Coefficient of thermal expansion

The coefficient of thermal expansion (CTE) is the linear change in dimension of the solder alloy when subjected to a temperature rise. As such, CTE is a thermomechanical property, but is listed under thermal properties for convenience. The CTE of solders (Table 2.11) is high, compared with the CTE of some of the substrates typically used for surface-mount applications (e.g., ceramic 3-5 ppm/°C, Kevlar-epoxy 6-10 ppm/°C, glass-epoxy 14-19 ppm/°C). The local CTE mismatch, as differentiated from the global CTE mismatch between surface-mount components and the substrate, leads to thermomechanical stresses in surface-mount solder-joint interfaces when they are subjected to variations in temperature during manufacture and use. Severe thermal excursions and repeated thermal cycling can result in overstress and/or fatigue failure of the solder interconnections.

2.4.4 Mechanical properties

The mechanical properties of solder are highly dependent on temperature, stress and strain range and rate, loading frequency, and time. An overview of mechanical properties of solders and solder joints is given in Lau and Rice [1985].

Constitutive behavior (stress-strain)

There is no single unified constitutive law that describes the deformation of solder alloys for the entire range of use conditions. However, some literature exists for limited conditions and for some alloys, especially isothermal deformation of eutectic alloys. Arrowood et al. [1991] reviewed the literature concerning constitutive models for tensile deformation, creep deformation, and stress relaxation of solder alloys.

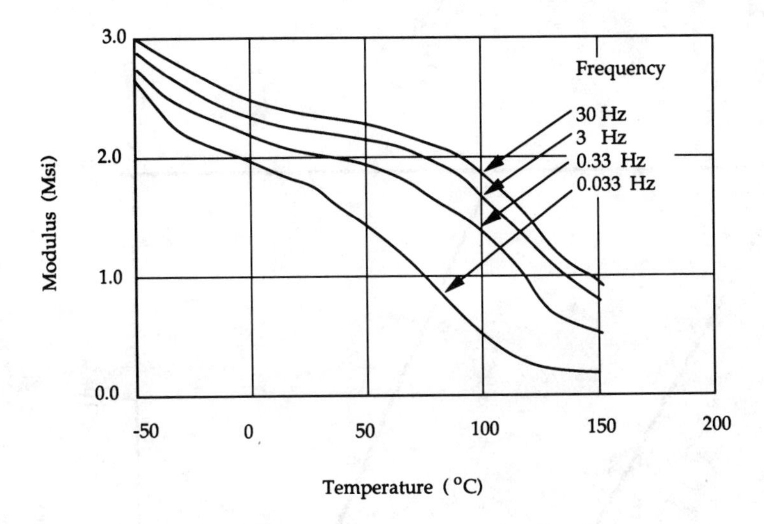

Figure 2.3: Modulus of 60Sn-40Pb solder as a function of temperature and frequency [Engelmaier, 1983]

Modulus, tensile strength, shear strength, and elongation

The modulus of solder alloys depends on the temperature and the rate of testing (Figure 2.3). The tensile strength of bulk solder and the shear strength of solder joints (ring and plug joints) have been measured as a function of the crosshead speed (stress ramping rate) and temperature. (Examples are given in Figure 2.4.) The strength increases with crosshead speed and decreases with temperature. Further, the size of the ring/plug joints does not significantly affect the shear strength per unit joint area. (The effect of thermal aging on joint shear strength is shown in Figure 2.5.) The percent elongation of solders increases with temperature [Ainsworth 1971].

Rupture strength and stress relaxation

The rupture strength is obtained by measuring the time to rupture of the solder joint when subjected to constant stress. The rupture strength of solder and solder joints decreases with temperature, as shown in Figure 2.6. The effect of thermal aging on rupture strength is shown in Figure 2.7. Stress relaxation of solders as a function of temperature has been studied by Baker [1979].

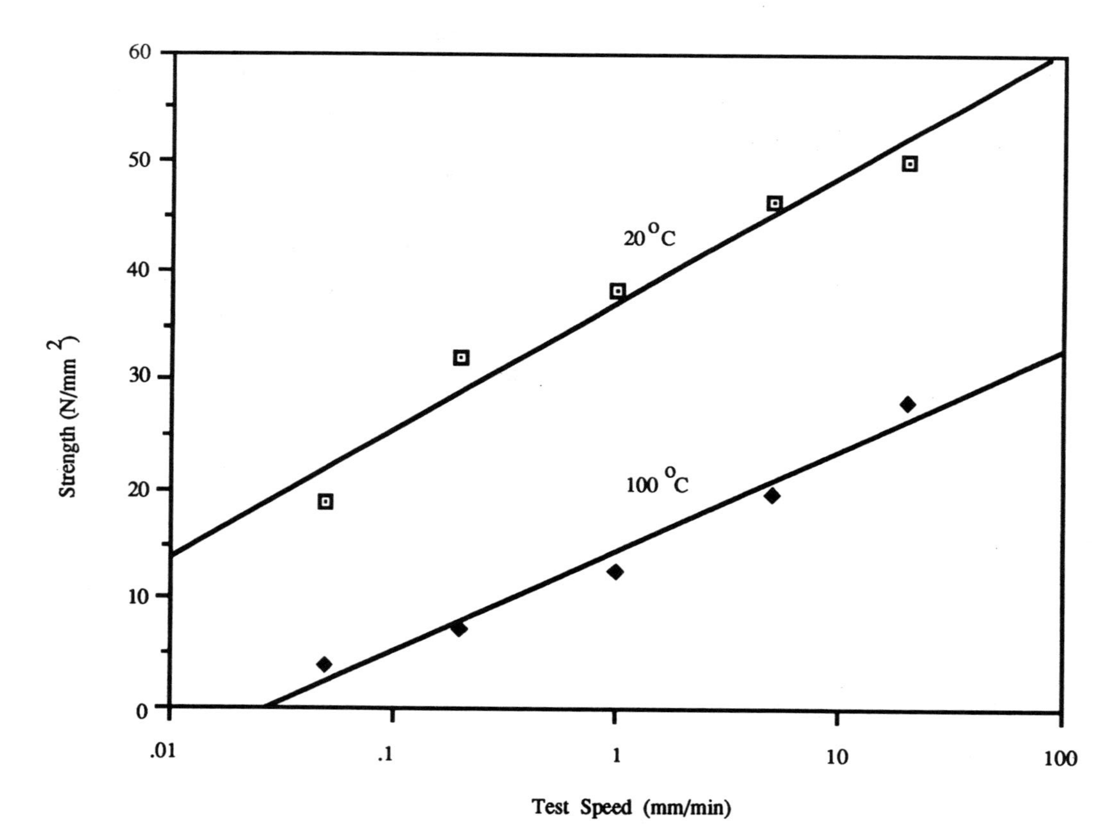

Figure 2.4: Tensile strength of bulk 60Sn-40Pb solder [ITRI Publication No. 656, 1987]

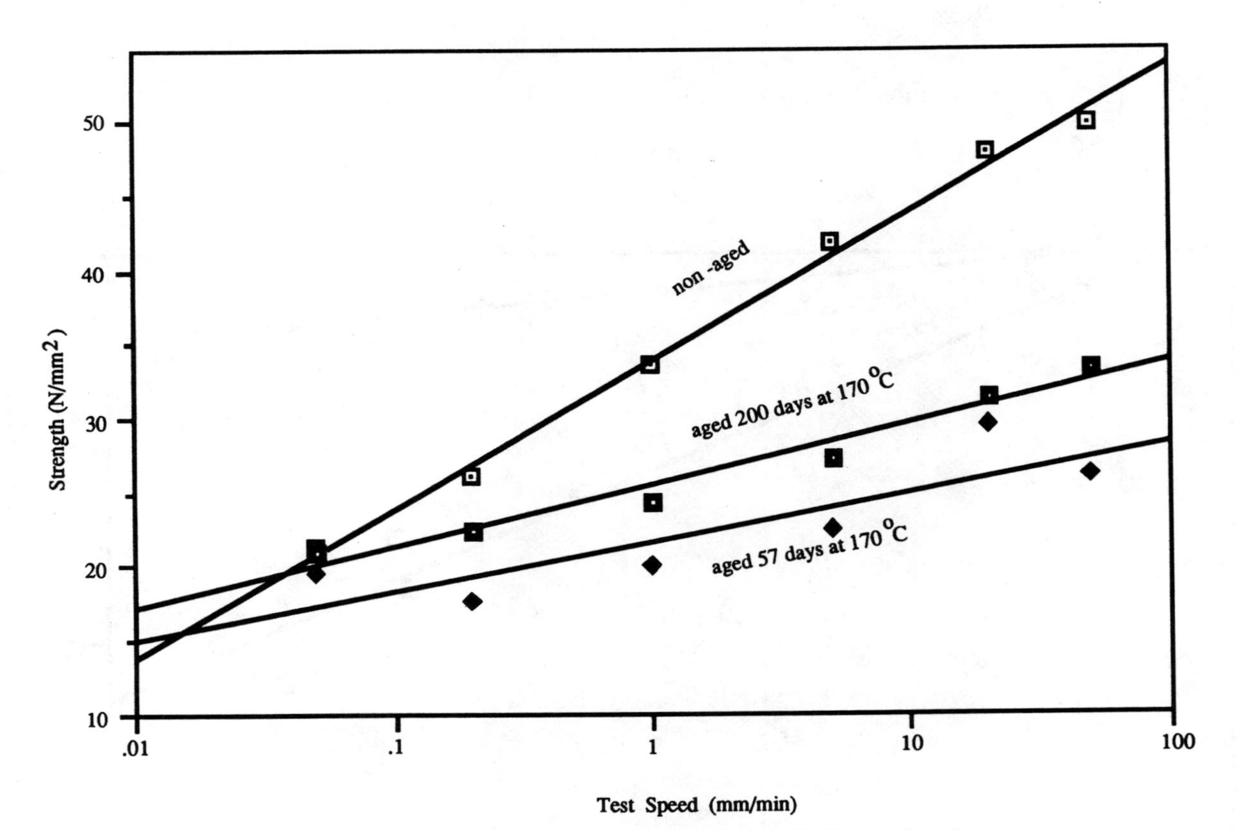

Figure 2.5: Effect of thermal aging on solder joint shear strength of 60Sn-40Pb solder joint at a test temperature of 20°C [ITRI Publication No. 656, 1987]

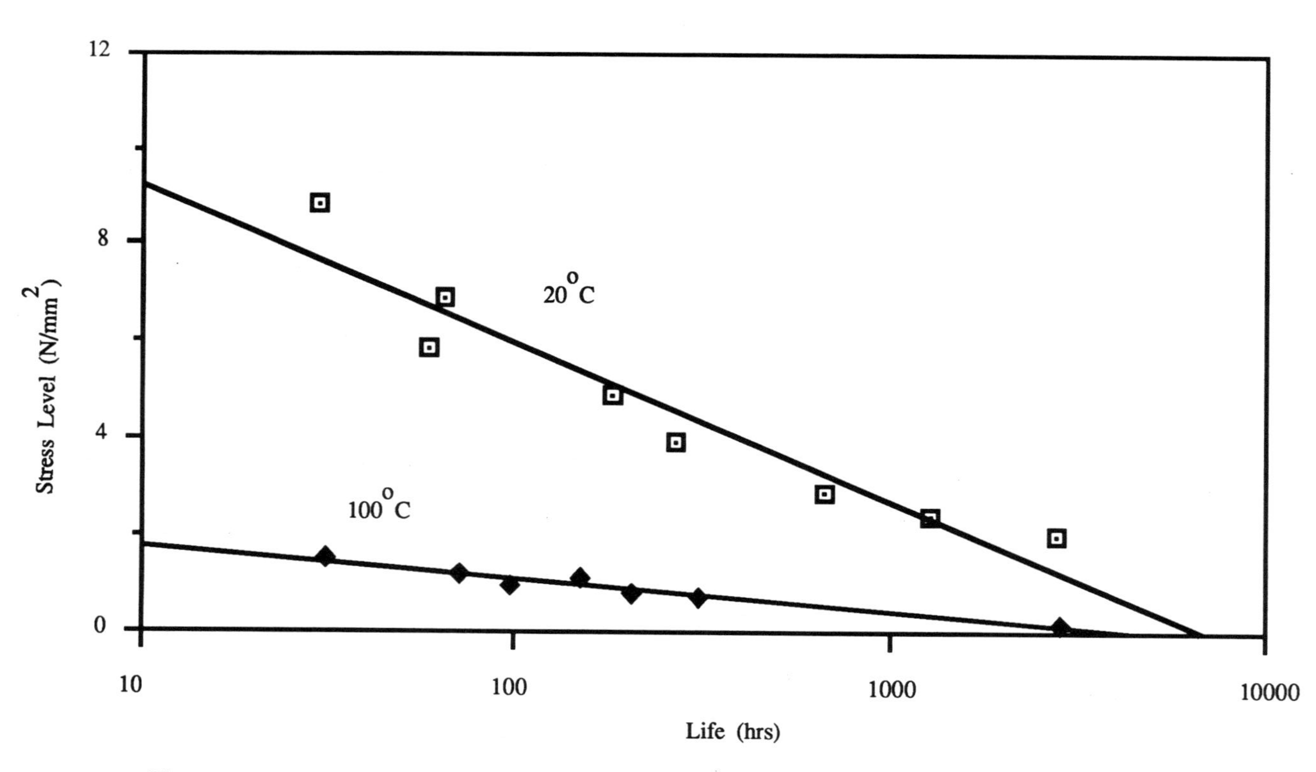

Figure 2.6: Creep rupture strength of bulk 60Sn-40Pb solder [ITRI publcn. no. 656 1987]

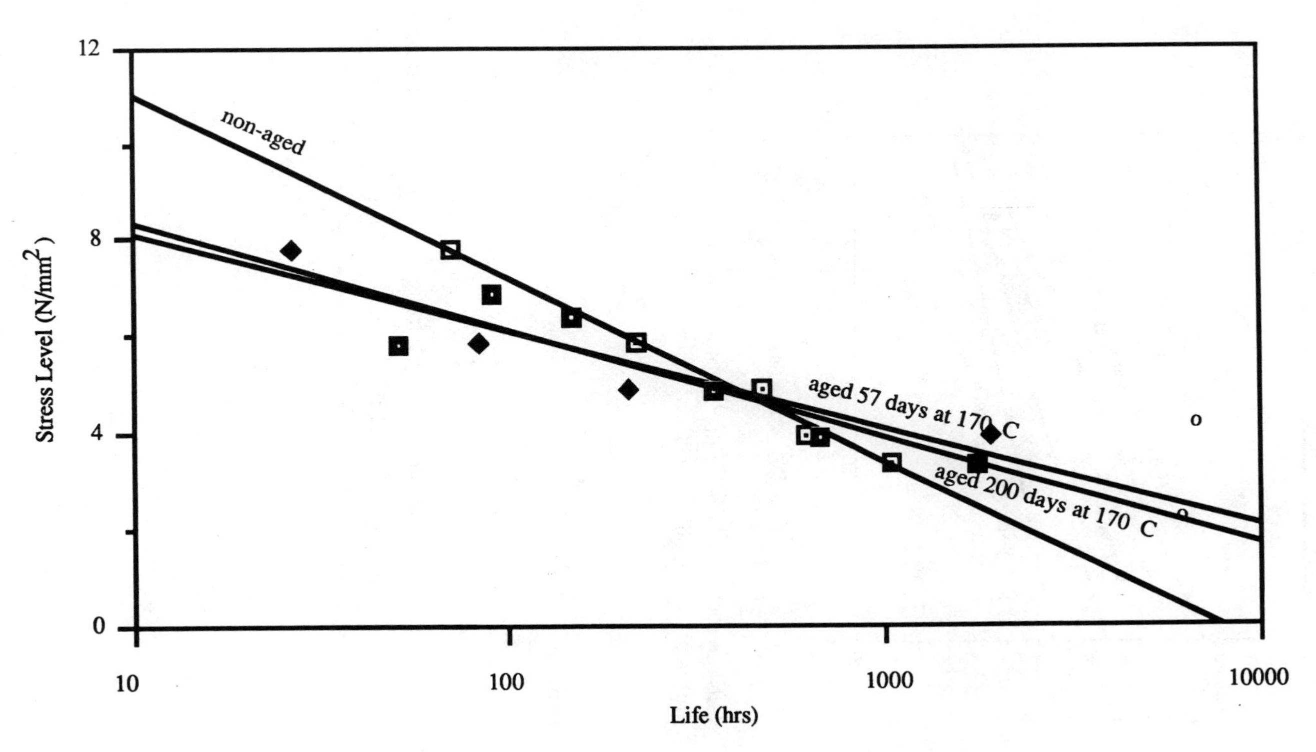

Figure 2.7: Effect of thermal aging on creep strength of 60Sn-40Pb solder joint at a test temperature of 20°C [ITRI publication No. 656, 1987]

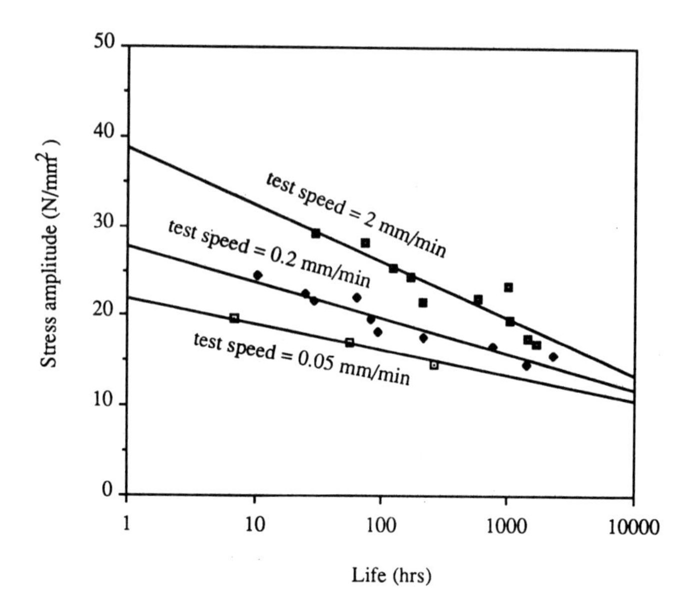

Figure 2.8: Fatigue strength of 60Sn-40Pb solder joint at a test temperature of 20°C [ITRI publication No. 656, 1987]

Fatigue strength

The fatigue strength of solder joints has been studied as a function of crosshead speed and temperature, as shown in Figure 2.8 and 2.9. Wild [1974] and Engelmaier [1983] have studied the effect of frequency and temperature on fatigue life for a given strain range (Figure 2.10). Vaynman et al. [1991] and Frear [1991] have extensively reviewed the literature on isothermal and thermomechanical fatigue of solders.

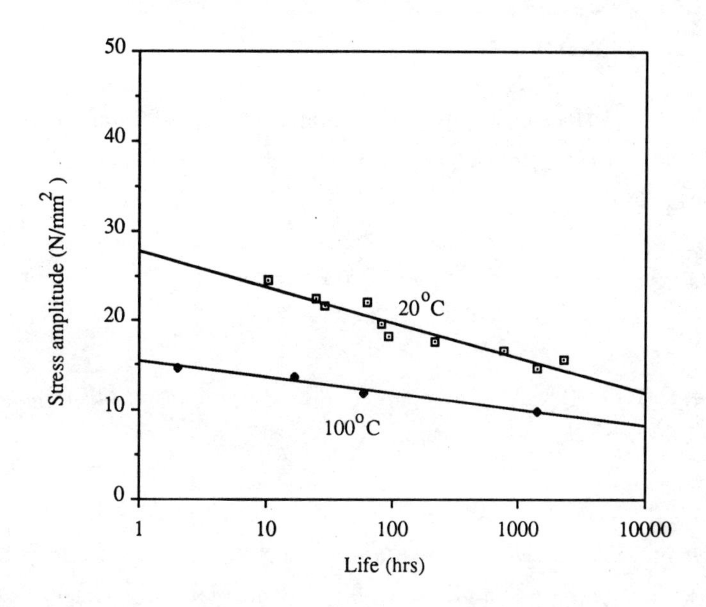

Figure 2.9: Fatigue strength of 60Sn-40Pb solder joint at a test speed of 0.2 mm/min [ITRI publication No. 656, 1987]

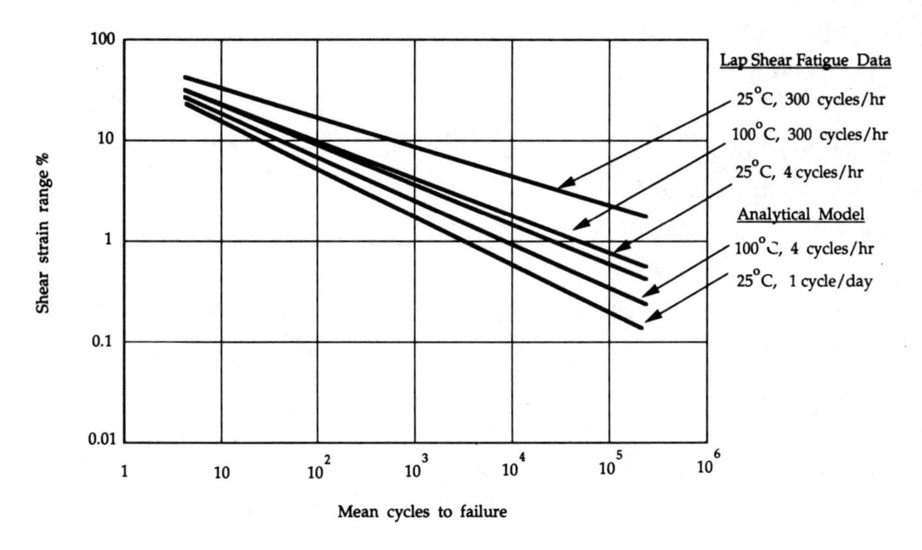

Figure 2.10: Fatigue properties of eutectic tin-lead solder

2.5 References

Ackroyd, M. L., Mackay, C. A., and Thwaites, C. J. Effect of Certain Impurity Elements on the Wetting Properties of 60% Tin-40% Lead Solders. *Metals Technology*, 2 (1975).

Ainsworth, P. A. The Formation and Properties of Soft Soldered Joints. *Metals and Materials* (1971) 374-379.

Andrade, E. N. C. The Viscosity of Liquids. *Proceedings of the Royal Society*, A-215, (1952).

Arrowood, R., Mukherjee, A., and Jones, W. B. Hot Deformation of Two-Phase Mixtures, in *Solder Mechanics, A State of the Art Assessment*, edited by Frear et al. Warrendale, PA: The Minerals, Metals and Materials Society (1991).

Baker, E. Stress Relaxation in Tin-Lead Solders. *Materials Science and Engineering*, 38 (1979), 241-247.

Becker, G. and Allen, B. M. The Effect of Impurities on Soft Solder for Electrical Purposes. *Proceedings Internepcon*, Brighton (1970).

CINDAS Report 102, 1991.

Engelmaier, W. Fatigue Life of Leadless Chip Carrier Solder Joints during Power Cycling. *IEEE Transactions on Components, Hybrids and Manufacturing Technology*, CHMT-6(3) (1983), 232-237.

Frear, D. R. Thermomechanical Fatigue in Solder Materials, in *Solder Mechanics, A State of the Art Assessment*, edited by Frear et al. Warrendale, PA: The Minerals, Metals and Materials Society (1991).

Hansen M. *Constitution of Binary Alloys.* New York: McGraw-Hill (1958).

Hwang, J. S. *Solder Paste in Electronics Packaging*, New York: Van Nostrand Reinhold (1989).

IPC-SP-819, General Requirements and Test Methods for Electronic Grade Solder Paste.

ITRI publication No. 656, 1987.

Lau, H. L., and Rice, D. W. Solder Joint Fatigue in Surface Mount Technology: State of the Art *Solid State Technology*, (October 1985), 91-104.

Manko, H. H. *Solders and Soldering*, New York: McGraw-Hill (1979).

MIL-F-14256, Flux, Soldering, Liquid (Rosin Base), Military Specification.

QQ-S-571, Federal Specification, Solder-Tin Alloy, Tin-Lead Alloy, and Lead Alloy,

Vaynman, S., Fine, M. E., and Jeannotte, D. A. Low-Cycle Isothermal Fatigue of Solder Materials, in *Solder Mechanics, A State of the Art Assessment*, edited by Frear et al. Warrendale, PA: The Minerals, Metals and Materials Society (1991).

Wild, R. N. Some Fatigue Properties of Solders and Solder Joints, NEPCON, (1974), 105-117.

Chapter 3

Wave Soldering

Anupam Malhotra
CALCE Electronic Packaging Research Center
University of Maryland
College Park, MD

Before wave soldering, printed circuit boards were soldered by either dip or drag soldering. Dip soldering involved placing the board, with the components on it, into a pot containing molten solder, taking care not to allow the solder to come over the top side of the board. Drag soldering, on the other hand, involved dragging the board over the molten solder for greater wetting.

The development of the wave-soldering technique resulted in a tremendous increase in productivity over the traditional dip and drag methods. The wave soldering process uses a wave of molten solder that is brought into contact with the printed circuit board in a solder pot. The board itself moves on conveyors across the top surface of the solder.

Key wave-soldering processes include fluxing, preheating, and soldering. Fluxing comprises the application of flux to the various assemblies while the preheater heats the board before soldering. It is very important for every joint to carry only the optimum amount of solder to produce an effective joint. Several methods have been developed for automated fluxing. Although the simplest and most preferred method is foam fluxing, there are other popular methods, including wave fluxing, spray fluxing, and brush fluxing.

Preheating is employed before actual wave soldering to raise the temperature of the board, activate the flux, reduce the number of blowholes in the solder joint, and reduce the thermal shock to the board from the solder wave.

Figure 3.1: A typical modern wave-soldering machine (*Electrovert*)

The boards, once fluxed and preheated, are soldered by a solder wave. The wave-soldering process is performed at a soldering station that includes the solder pot, the solder pump and the wave-forming nozzle. The boards are transported through the various parts of the machine on conveyors.

Modern wave-soldering machinery comes with a variety of equipment to enhance productivity and efficiency including dynamic solder pots; inclined conveyors; hot air and quartz preheaters; oil injection; computer control; machine vision; automatic placement equipment; and air knives [Lambert 1988].

Wave soldering depends on many parameters that affect the overall performance of the soldering system: flux density, foam head, solder alloy, solder temperature, conveyor angle, conveyor parallelism, oil adjustment, dross level, solder wave, and smoothness of the wave.

3.1 Fluxing

Fluxing chemically cleans the surfaces to be joined, cleans the surface of the solder powder, and maintains the cleanliness of both substrate surface and solder-powder surface [Hwang 1989]. The process consists of applying a flux (a chemical compound with alcoholic and acidic contents that react with the flux container) to the solder-joint locations before soldering. Flux is stored outside the wave-soldering system and is continuously circulated through the fluxer to prevent separation or stratification. A stable temperature is required to maintain the effectiveness of the flux. Flux is stored in a polypropylene container, but the alcohols in the flux can adversely react with plastics, causing them to become brittle. Therefore, containers made of metals that do not react with the acids in the flux are preferred.

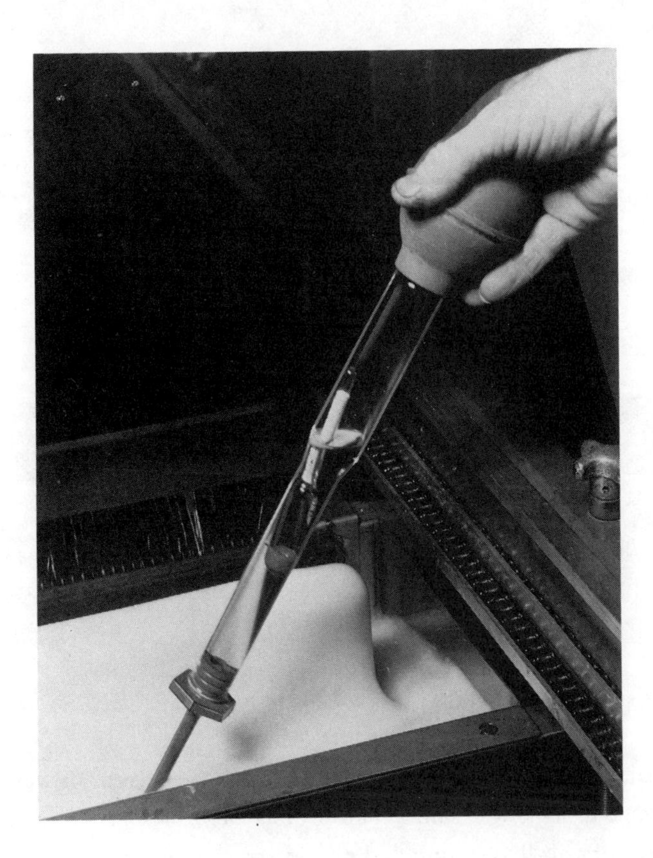

Figure 3.2: Checking the specific gravity of foam in a foam fluxer (*Electrovert*)

The flux container holds a porous air stone used to generate a foam head. Pressurized air is sent through the stone, and escapes through its pores, causing the foaming action. If the flux reacts with the air stone its characteristics will change and overall soldering performance will be affected.

The fluxing process requires careful control of the density or specific gravity of the flux. Manual methods can be used, as shown in Figure 3.2, although hydrometers or automatic viscosity measurement systems are more accurate and faster. The temperature of the flux solution is another important parameter requiring regulation.

The following sections discuss the various methods of applying flux used in the electronics industry.

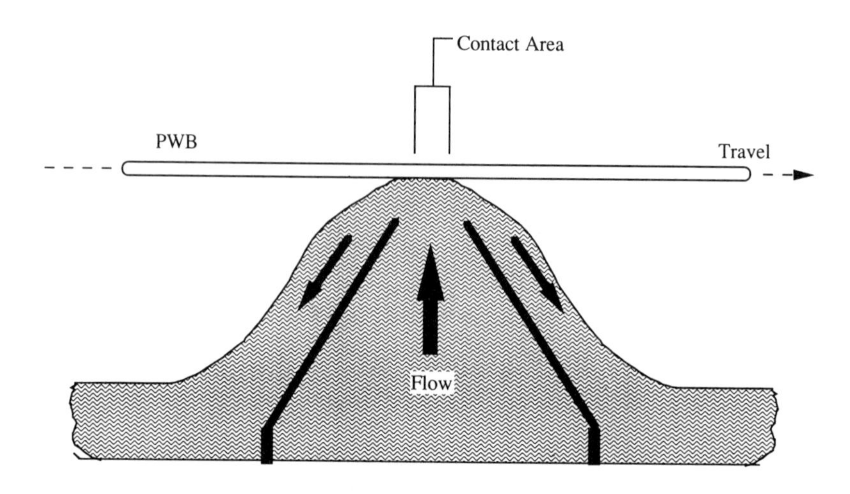

Figure 3.3: A parabolic wave [Lambert 1988]

3.1.1 Wave fluxing

The soldering process was greatly simplified with the advent of wave-soldering technology, and soon the same concept was being applied to fluxing. The wave fluxer is basically a pump-and-nozzle arrangement mounted in a tank or container. The pump forces the liquid flux up through a nozzle, from which it projects upwards and falls back into the container, forming a parabolic wave (see Figure 3.3). The board passes over the wave so that the underside just touches the tip of the flux wave, spreading the flux over the underside of the board and helping it stick to the component leads to be soldered.

The wave fluxer generally works at high wave heights, providing a smooth wave that avoids getting flux on the non-flux side of the board. The wave is smoothed by putting a wire mesh or perforated screen over the nozzle to reduce ripples and to avoid wetting unfluxed areas.

Operation and setup of the wave fluxer is not as critical as it is for other fluxers. However, to reach zero-defect soldering, the maintenance and operation of the fluxing equipment must be properly controlled, particularly because it is difficult to determine when the wave fluxer is working within tolerances. On the other hand, for example, foam fluxing indicates proper operation by the characteristics of the foam head.

Typically, an air knife is provided at the fluxing station to assist the flux's return to the tank. The air knife produces a flat stream of compressed air that hits the underside of the board at an angle of about 5° towards the wave, causing any excess flux to go with the stream of flux and fall back into the fluxer. Otherwise, the excess flux on the board can drip into the preheater coils, igniting flammable components in the flux and causing a fire.

3.1.2 Foam fluxing

In foam fluxing, the flux produces a stable head of foam when air is forced through the liquid. The foaming action depends on the type of flux used and any detergents added to the flux for the purpose of cleaning away dirt and grease from the solder site. Foam fluxing requires a minimum setup time and minimum maintenance.

The foam is produced by blowing compressed air into the flux; this air passes through a porous tube from either one or both ends. The porosity of the tube allows extremely fine bubbles of air to pass through to the flux, causing the flux to rise and form foam. As more air rushes in from the tube, the foam is pushed away from the outside of the tube toward a nozzle. The foam ultimately rises up to the nose of the nozzle and protrudes to a height of about 1/4 in. (6 mm). The foam produced has a constant head, and its height above the nozzle is adjusted to prevent the foam from coating the top side of the board as it passes over the nozzle. The excess foam then flows down the side of the nozzle into the foam fluxer, preventing waste.

Typically, an air knife is provided to help the foam return to the fluxer. Or, instead, a brush may wipe the excess flux back into the fluxer. However, the brush tends to get clogged with flux after some time and requires cleaning. Moreover, if the bristles are too stiff, the components may shift from their position; if they are too soft, the components are likely to get soaked with flux.

Maintaining a sufficiently high head of foam to completely wet the protruding leads is difficult in foam fluxers. Long leads are more difficult to wet, because a foam fluxer can typically provide only about 3/4 in. (10 mm) of foam above the nozzle top with a standard nozzle. If a greater head is required, brushes are used to support the nozzle, to help the foam reach the board, and to allow the longer leads to pass through their bristles. Because stiff bristles tend to displace the components on the board, extender wings are sometimes used to extend the foam, but they cannot provide a head of more than 12 mm.

Although the foam fluxer is a simple piece of equipment, specific requirements must be met for proper maintenance and operation.

- The air supply to the porous tube or stone must be increased slowly to avoid forming bubbles. Small bubbles form a fine foam, producing the best results. However, for more foam, more air is required. More air causes bigger bubbles, which tend to burst and cause the head of the foam to vary. To avoid this phenomenon, the air supply is increased only up to the point when larger bubbles begin to form.

- The foam head must be stable. If there is no way to obtain a satisfactory head of foam, the flux must be checked for contamination. Contaminated flux must be disposed of, and the fluxer must be cleaned thoroughly with a solvent and washed in hot water.

- Dry flux must be avoided, because it is the primary cause of blockage of the air stone. If the fluxer has to be shut down for a long period, the stone is stored in a thinner solution in a suitable container.

- All non-operating fluxing equipment must be cleaned and washed with a thinner. Flux is a sticky material and will adhere to the container if allowed to dry.

- The pallet must never be allowed to enter the fluxing section while hot; otherwise the foam head will collapse.

- The compressed air must be clean and oil-free. Any contaminant will prevent the flux from forming a good head.

3.1.3 Spray fluxing

Spray fluxers are the messiest fluxers because they deposit the flux all around the target site. Spray fluxers should be avoided unless specifically needed, as in the case of fluxing components with long leads that tend to interfere with other fluxers. The various types of spray fluxers available are discussed in the following paragraphs.

Compressed air spray fluxer

The compressed air spray fluxer uses the technology of spray painting — sending a blast of compressed air through liquid or powdered flux to force it out the nozzle as a fine spray. Because the flux contains corrosive elements such as acids and alkalies, the container and the nozzle are made of corrosion-resistant materials.

Controls ensure repeatability of the process and protect against overspray. These systems are generally only manufactured in-house and are not commercially available. A major drawback of these systems is that the flux dries out and subsequently blocks the nozzle. Moreover, used flux is not recoverable.

Airless spray fluxer

The airless spray fluxer, shown in Figure 3.4, avoids the problem of contamination in the flux air and prevents nozzle blockage. The flux itself is pressurized and forced through the nozzle from which it emerges as a finely atomized mist of flux particles. The advantage of this method is that it prevents the nozzle from becoming blocked. However, the airless spray is still generally used as an in-house manufactured piece of equipment, rather than a commercially available product.

Drum and air knife spray fluxer

The drum and air knife spray fluxing technique does not produce a spray in the strictest sense of the word. Rather, a flow of droplets is formed and deposited on the underside of the board. The equipment consists of a revolving drum with perforations in the form of a mesh. The drum is rotated slowly by a motor and has an air knife inside its periphery that projects air towards the underside of the board. As the drum rotates, the flux is picked up from the bottom of the container in which the drum resides and sticks to the pores in the drum. The air knife blows a stream of air toward the perforated surface, causing the flux to be blown onto the board in the form of droplets. Both the speed of the air through the air knife and the speed of the drum can be controlled to vary the volume of flux deposited on the board. The only disadvantage is that the contaminated flux drops back into the container without having gone through a purification stage first. Even though the amount of contamination does not constitute a major problem, today's emphasis on zero-defect soldering requires all possible precautions.

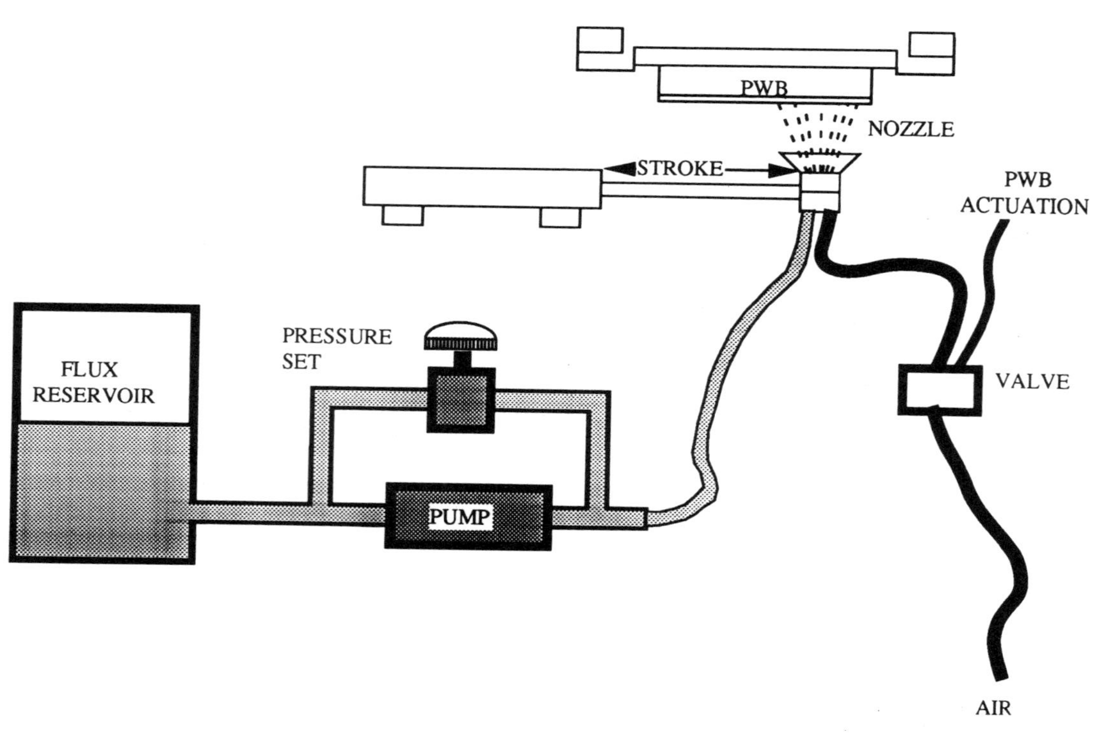

Figure 3.4: Airless spray fluxer

Nozzle-droplet spray fluxer

The nozzle-droplet spray fluxer has a primary compressed air nozzle surrounded by secondary flux nozzles. Compressed air is passed through the primary nozzle, sucking in the flux from the secondary nozzles to form a fan-shaped spray of flux droplets. The board gets sprayed on the underside, while the rush of compressed air pulls in more flux from the secondary nozzles and continues the process.

Because the flux does not mix with the air, the problem of flux drying, contaminating, and blocking the nozzle cavities is averted. The excess spray falls back into the container in the form of droplets. The flux is not contaminated since, it does not get into contact with air.

All nozzles are lined up and can be individually controlled to produce the flux-spray pattern desired for the type of board being fluxed. There is much less overspray in the nozzle-droplet spray fluxer than in other forms of spray fluxing. However, the system must still be properly maintained and cleaned for trouble-free operation.

3.1.4 Brush fluxing

Brush fluxing, rarely used today, is simple in design and operation. A revolving cylinder with bristles attached to its surface applies the flux onto the board as it passes over the container. The brush revolves between the flux in the container and the underside of the board, lapping up flux from the container to the board. However, components may become displaced due to the stiffness of the bristles. Moreover, there is no control over the amount of flux being applied to the board; thus, if components are the through-hole type, the inside surfaces of the through-holes in the board may not be properly wetted.

The brush fluxer has a recirculating system and, therefore, the density and purity levels of the flux being applied must be checked. Because the bristles tend to brush away oil stains, fingerprints, and grease from the board, contaminating the flux, the brush must be cleaned regularly.

3.1.5 Dip fluxing

Dip fluxing involves placing the printed circuit board on the surface of a solder bath. The top of the container is generally kept open, so extreme care has to be taken to avoid contaminating the fluxing solution and to prevent undue evaporation of solvent from the flux. Automation is difficult, because the board must actually be dipped into the flux without covering the top surface of the board.

3.2 Preheating

The purpose of preheating is to accelerate the soldering process by increasing the temperature of the board before passing it through the solder wave. Preheating also reduces the thermal shock from the solder wave on the printed circuit board, activates the flux by providing heat, and evaporates the flux solvents, which, if evaporated during solder-joint formation, can cause blowholes in the solder joint.

Printed circuit boards are usually designed to withstand a 10-sec. float on a solder bath without any damage to the laminate or cracking of the copper. The assumption is that any board that can tolerate such a severe endurance test can hardly be affected by a brief passage through the solder wave. Moreover, only the leads of leaded components come into contact with the solder wave. In terms of thermal shock to components, the possible damage to their body due to a passage through the solder wave is reduced by including a preheat process. Preheating experiments have shown that a board soldered at 12 ft. (4 m) per min using the normal soldering process can be soldered at 2.5 ft. (0.7 m) per min with preheating [Woodgate 1988].

The more densely packed the components are on the board, the more preheating is required, because the components absorb much of the heat transferred during the soldering process. The copper embedded between the layers of multilayer boards also tends to remove heat. To compensate for this loss, top as well as bottom preheaters are used, and heat is also supplied by the solder wave. However, the components should not be exposed to the solder wave for too long, or the interconnections and circuitry within the components may be degraded.

Preheaters are generally classified as electrically heated (including hot-plate, platen, and cal-rod preheaters), air-heated, and infrared-light-heated (including quartz-plate heated, quartz-rod heated and infrared-lamp heated). An ideal preheater:

- has the ability to quickly heat the metal in the joint without affecting the temperature of the laminate,

- is energy efficient,
- is self-cleaning,
- has even heating across the board,
- has an adjustable heating surface to accommodate different jobs, and
- has an unvarying thermal output.

The actual performance of various preheaters does not vary much, so selecting a preheater generally depends on the type of work to be performed and other factors, such as available space and cost considerations. Only after testing to determine the performance of the preheater under the severest possible conditions is a decision made. On the other hand, often the simplest device that requires minimum maintenance and still gives the required results is selected. The following sections, describe various types of preheaters.

3.2.1 Electric preheaters

Hot-plate preheaters

The hot-plate preheater, a flat plate heated by strip heaters, is the simplest preheating device. The plate is typically about 1/4 in. (6 mm) thick. To conserve energy, multiple strip heaters are aligned parallel to the line of travel of the boards and some of the strips are switched off if a narrower board passes through. Heat losses at the edge of the plate are avoided by providing separately controlled edge heaters.

The time required for initial heating of the hot plate can be as great as 45 minutes. Therefore, hot-plate preheaters are generally used for soldering operations involving the running of several boards at the same temperature. The boards should never be placed on aluminum foil while preheating, because the foil reflects the heat from the preheater coils, harming the surface of the heater and reducing its effectiveness. The hot plate is self-cleaning, because any residues left on its surface are usually burnt off by the heat of the plate, although some remnants must be occasionally scraped off with a spatula.

Cal-rod preheaters

The cal-rod preheater uses tubes that reflect the infrared light emitted during operation at red heat (i.e., the temperature at which the cal-rods glow red).

Cal-rod tubes are fixed in a frame and aligned parallel to the incoming boards; changes in the size of the board are accommodated for by switching the cal-rods on or off. The amount of heat reflected varies with the cleanliness of the reflecting surface. To avoid constant cleaning, the reflecting surface is made by covering a non-reflecting surface with a sheet of aluminum kitchen foil that is replaced when dirty. The cal-rod system heats up faster than the hot plate, but is about the same in energy efficiency. The cal-rod preheater requires minimal maintenance, because it operates at red heat and is self-cleaning.

3.2.2 Air preheaters

Hot air is a good choice for heating boards when the flux uses alcohol rosin, because hot air removes flammable vapors from the board. The hot air preheater is generally used in combination with some of the other types of preheating, like plate or cal rod heating, because of the ineffectiveness of convective heat transfer.

3.2.3 Infrared preheaters

Quartz-plate preheaters

The quartz-plate preheater consists of a thick quartz plate on which the heating elements are embedded in slots or pockets. The energy from these preheaters is in the $3-\mu m$ bandwidth of infrared heat; there is also some convective heating. Thermocouples are enclosed to monitor the temperature of the preheater.

Quartz-plate preheaters are fastest in terms of heating and cooling rates (about 93°C/min.). They are self-cleaning, heat evenly, and can accommodate boards with different widths. Among the disadvantages are high cost, fragility at high temperatures, and the complex control circuitry required to manage the preheater's quick response time.

Quartz-tube preheaters

The quartz-tube preheater is similar to the quartz-plate preheater, except that it uses tubes, rather than a plate (as in the cal-rod preheater). The quartz-tube system is faster than any of the other preheating methods. Moreover, not only can the rods be placed parallel to the board travel to adjust for most board widths, but the system can also work with the rods placed across the conveyor to control the amount of preheat. The greatest advantage, however, is that the quartz-tube preheater provides heating not only by infrared heating, but also by blowing air through the tubes. The required air is input from the immediate vicinity of the preheater and heated up by the tubes. These advantages override cost and fragility considerations for large-scale manufacturing.

Lamp-type infrared preheaters

Where fast heating, preheating, and high temperatures are required, the lamp-type infrared system is the best method available. The trade-off, however, is that great control over the lamp is required to avoid excessive heat damage to the components or the board. Moreover, the lamps are costly and their gold-plated reflectors need water-cooling jackets. Dust protection is provided for these devices because attempting to wipe dirt off can result in damage. Despite their cost, such infrared systems are quite common in automatic soldering equipment.

3.3 The solder wave

The most important parameter to be controlled in a solder machine is the solder wave. The nature of the wave determines the effectiveness of the soldering operation. Figure 3.5 classifies the solder waves discussed in this section. In addition both single solder wave and the dual solder wave processes will be presented.

3.3.1 The symmetrical wave

The geometrically symmetrical wave is shown in Figure 3.6. The symmetrical wave is still used, although its width and the nature of its profile are often modified. The modifications help increase the effectiveness of the symmetrical wave by increasing the contact area of the wave on the board and increasing the force with which the wave impacts the board. This, reduces the incidence of defects in the solder and increases the productivity of the process.

The primitive symmetrical wave

The primitive symmetrical wave was the first solder waveform developed for surface-mount components. Earlier forms of the wave touched the board at its

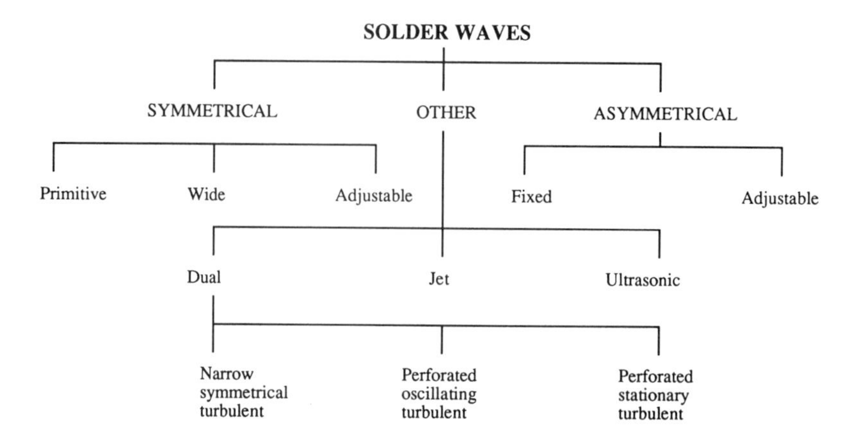

Figure 3.5: Classification of solder waves

crest, with the board moving horizontally. The consequent high relative motion between the board and the wave at the point of exit caused the solder to cool improperly and allowed insufficient time for the excess solder to drain back into the bath. This problem was solved by using a small amount of oil in the solder, usually in the form of a thin film on top of the wave, to reduce the surface tension between the solder and the board and to reduce the surface oxidation of the solder.

Wave-soldering machines use a symmetrical wave with the board moving at an angle of 2° to 7° to the horizontal (Figure 3.6). The topmost part of the wave is a region of slow-moving solder; therefore, as the board exits from this point, the speed of cooling decreases. In contrast, at the point of incidence, the solder wave flows toward the board, properly wetting the underside. The surface tension of the solder helps tack the excess solder back to the bath.

The wide symmetrical wave

With a solder wave, an angled conveyor is employed, although for simple boards a horizontal conveyor can be used, especially when soldering high volumes. The wide symmetrical wave is produced with nozzles that open as far as practical. The limit of opening for the nozzle is dictated by the pumping force available and the wave height desired. However, the nozzle can be widened with support or extender plates.

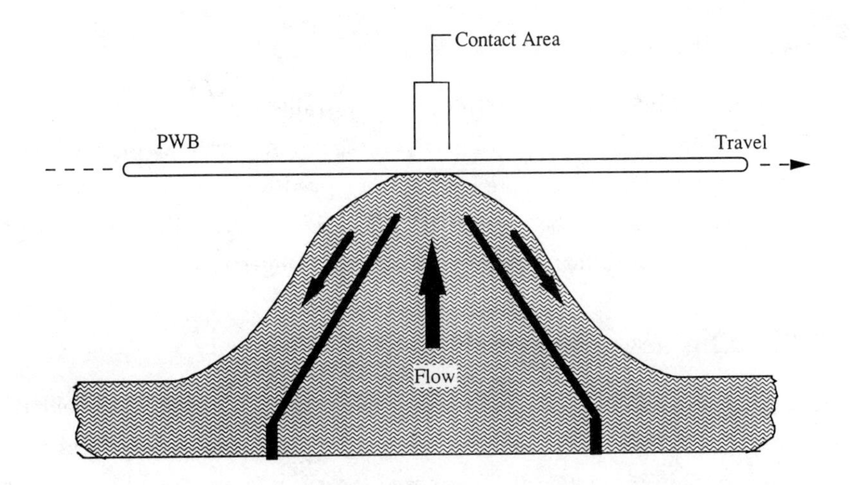

Figure 3.6: Parabolic wave with angled conveyor

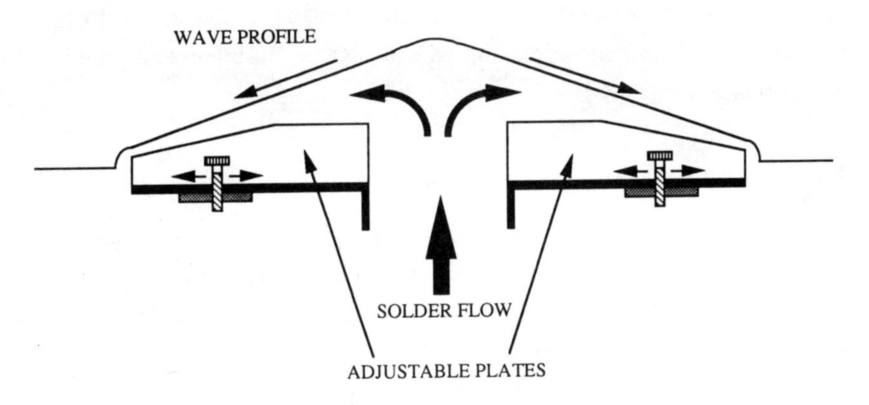

Figure 3.7: An adjustable symmetrical wave

The adjustable symmetrical wave

An adjustable wave (see Figure 3.7) uses adjustable extender plates with the nozzle to widen the wave area. The advantage gained is that high soldering speeds can be attained for simple boards when used with horizontal conveyors. Widely spaced plates can be used to get a wider wave, with the central region of the wave being used for soldering boards with longer leads.

3.3.2 The asymmetrical wave

The asymmetrical solder wave reflected developments in earlier wave-soldering methods. The purpose was to improve wetting and the removal of excess solder from the printed circuit board. The wave form is asymmetrically shaped by using the different techniques discussed below.

The fixed asymmetrical wave

The fixed asymmetrical wave is essentially a version of the wide symmetrical wave that uses an extender plate on only one side of the nozzle to improve contact area and provide a less abrupt exit region. Both horizontal and angled conveyors are associated with its use.

The adjustable asymmetrical wave

The adjustable asymmetrical wave is the most effective wave form used in wave soldering. It is designed to confine the functions of heating and wetting of the solder joint and the removal of surplus solder to two different portions of the wave. (Figure 3.8 is a schematic of such a wave.) The conveyor approaches the wave at a very small angle and meets the majority of the molten metal flowing in the opposite direction, as shown in the figure. This fast-flowing stream of solder heats up the component lead ends and wets them to form a soldered connection. The small volume of solder flowing in the opposite direction (along the horizontal adjustable exit wing in Figure 3.8) forms a static lake of molten solder. As the board moves further up into the wave, it reaches this horizontal, static region. At this point, the board separates from the wave and the high surface tension of the solder, along with the small angle of separation between the horizontal wave surface and the slightly inclined conveyor, drains away the excess solder. The disadvantage of this process is that setting up nozzles

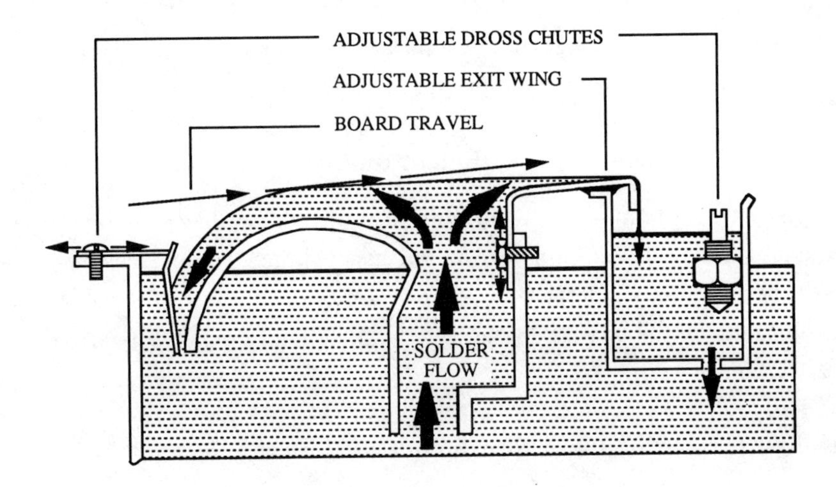

Figure 3.8: An adjustable asymmetrical wave

for optimum performance is difficult, since the height of the wave cannot be adjusted without jacking up the solder pot.

3.3.3 The dual wave

A combination of a narrow turbulent symmetric wave and a slow-moving, wide, horizontal wave, known as a dual wave, overcomes some of the limitations of the other wave forms. The concept involves the use of a flat moving stream of solder to scrub the surface of the assembly, while a steady horizontal solder lake removes the excessive solder from the board. (Three different kinds of dual wave systems in use are shown in Figure 3.9.)

The main problem solved with a dual wave system is the shadow effect of surface-mount components that are not aligned in the direction of the wave [Capillo 1990]. Components that are aligned perpendicular to the direction of motion of the printed circuit board are easily wetted on the side facing the wave. However, the component body tends to pull the wave toward the edge parallel to the direction of motion. Therefore, the lead terminations on the opposite side of the component, which are against the direction of motion of the printed circuit board, come under the shadow of the component and are not wetted by the solder wave. Taller components pose a greater problem because their shadow extends a greater distance from the component boundary; the length of the leads has to be increased to overcome this effect.

The narrow symmetrical turbulent dual wave

In this wave system, a narrow symmetrical wave is produced by liquid solder emerging from a slotted chimney. An adjacent chimney produces a slow-moving asymmetrical wave. The latter wave overcomes the shadow effect of the large components on the board and rectifies the soldering errors of the narrower and less turbulent wave.

The perforated oscillating turbulent dual wave

In the perforated oscillating turbulent dual wave system, an oscillating cylinder with holes perforated at regular intervals on its periphery provides the primary wave. This wave is usually a stream of bubbles of molten solder extruding from the holes in the cylinder. The cylinder oscillates to provide turbulence to the wave to increase its scrubbing action. Both the speed of oscillation and the solder flow rate can be varied to provide the required scrubbing action for a particular board.

The perforated stationary turbulent dual wave

The perforated stationary turbulent dual wave has a stationary perforated cylinder acting as the nozzle for ejecting the solder wave. Because the nozzle is stationary in this system, the turbulence is much less than in the movable cylinder type of wave system. A hot air knife is used to eliminate solder bridges and shorts.

The jet wave

The jet wave, depicted in Figure 3.10, is unique in that it has no turbulent zone to wet the component leads. Instead, a high-velocity molten solder wave jet is impinged upon the board underside. The jet creates a wave that is hollow beneath the surface; the hydraulic force exerted on the board is therefore much higher than in any other system. Moreover, the board's entry into the hollow part of the wave creates a turbulence that helps wet the leads with solder.

Wave soldering can be used successfully on conventional through-hole assemblies. However, the problem of solder bridging increases for packages with

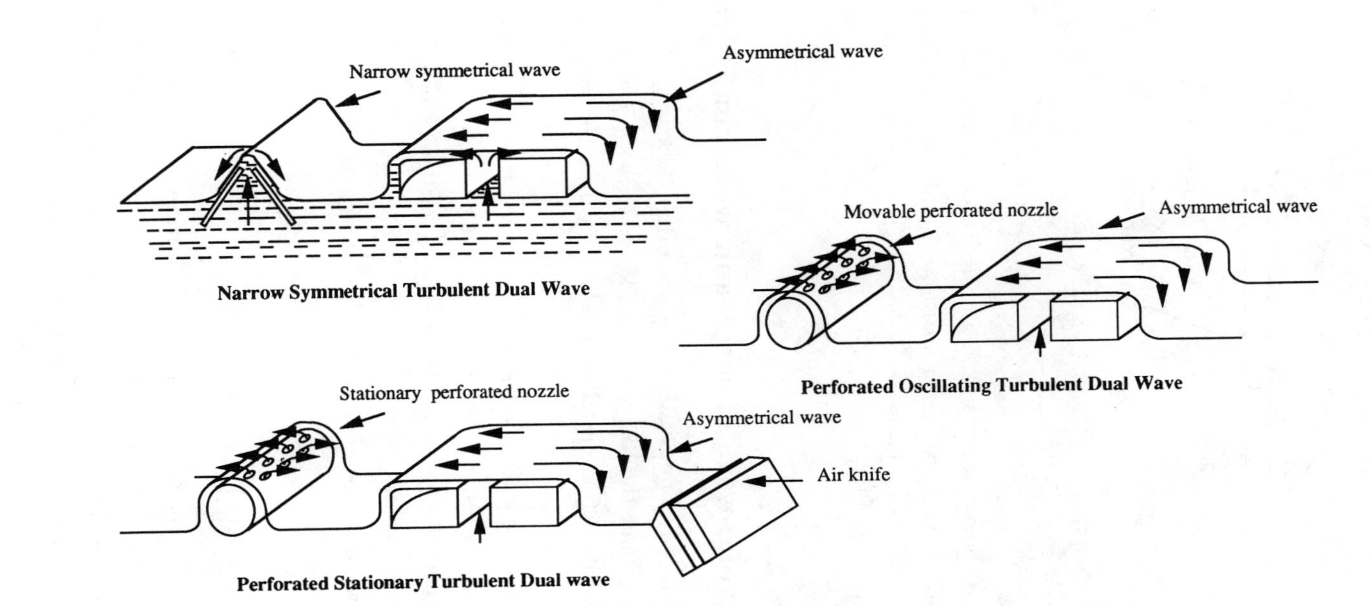

Figure 3.9: Various forms of the dual wave [Capillo 1990]

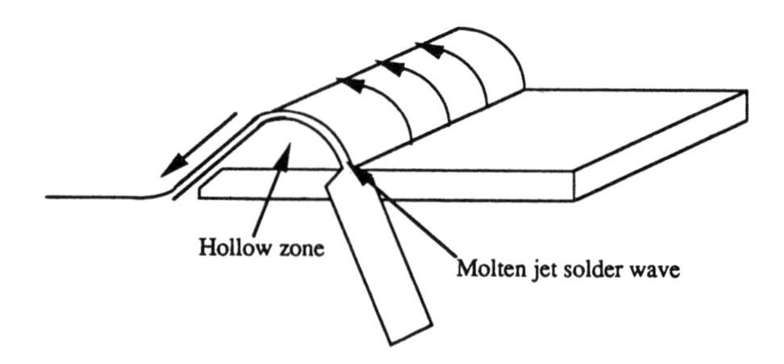

Figure 3.10: A jet wave [Capillo 1990]

lead pitches at or below 50 mils (1.27 mm). Hence, wave-soldering systems are predominantly used for soldering boards with underside attachments. The wave-soldering process can, however, be used with other processes to produce mixed-technology assemblies.

3.3.4 Dynamic wave concept and wave configuration

The solder wave is a dynamic wave in the sense that there is a relative motion between the board and the solder wave. The solder wave can be configured geometrically and its action described mathematically. The movement of the solder generates a multitude of forces that increase its wetting action and also create points where the solder flow is preferentially directed so as to allow the drainage of excess solder into the solder bath, thus preventing the formation of solder icicles. (Figure 3.11 shows the cross-section of the wave and depicts the various forces acting on it.)

The solder wave can be divided into three zones, as shown in Figure 3.12 and discussed below.

Zone 1 (point of entry)

The first zone is a turbulent part of the solder wave. The solder leaving the solder nozzle strikes the board and forms a "waterfall," which then falls back into the sump. The goal is to induce a large enough relative velocity between the board and the solder wave to ensure proper wetting of the component leads, as well as a washing action to cleanse the board of excess flux that may hinder effective joint formation.

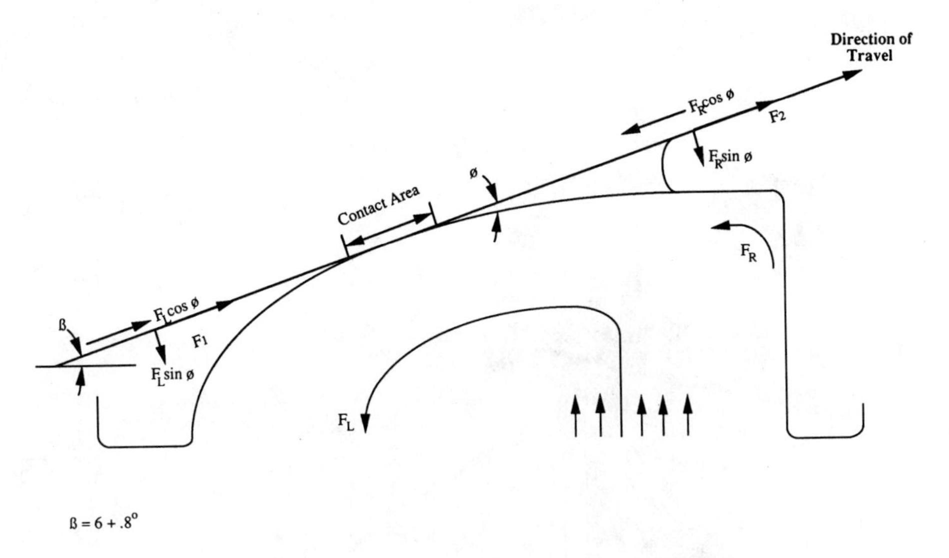

Figure 3.11: Solder wave configuration and analysis

Because there is only a laminar flow of solder along the board in Zone 1, there is not enough turbulence to force the removal of the sticky flux from the laminate between the conductors, though it is generally removed from the metallic lands.

The wetting mechanism depends on the coating applied to the metal before soldering. Three types of coatings are generally applied to the base metal:

- **Fusible coatings.** These are generally made of tin and lead and do not contaminate the pot. They melt in the heat of the solder wave and leave a clean surface of base metal to wet.

- **Soluble coatings.** These coatings, generally gold or silver, dissolve in the solder bath. Due to the fast processes involved, the removal is generally not complete and traces of silver and gold remain on the surface of the board. These traces become impurities in the solder and must be removed before excess solder returns to the solder pot.

- **Stable coatings.** These coatings neither dissolve nor melt at the soldering temperature, and do not contaminate the solder in the pot.

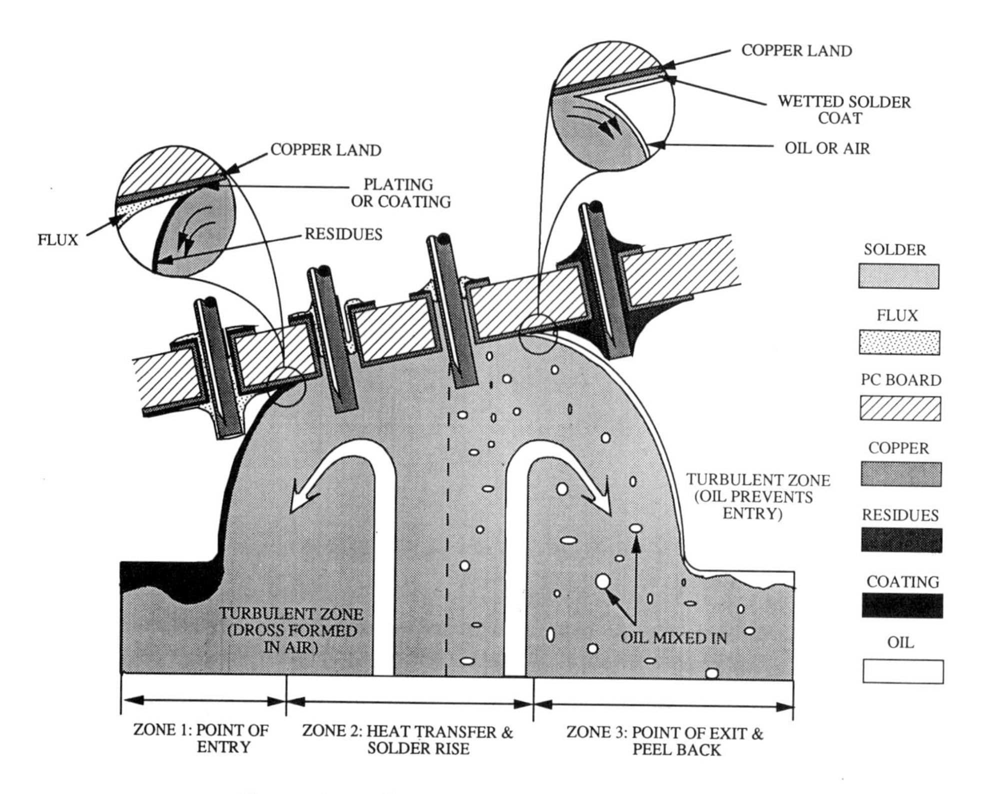

Figure 3.12: Solder wave zones [Manko 1992]

Zone 2 (heat transfer and solder rise zone)

The second zone is the solder wave. As shown in Figure 3.12, the board spends most of its soldering time in this zone to allow the solder to properly rise into the through-holes and completely wet the component leads.

The hydrostatic pressure of the solder from the nozzle helps wet the underside of the board and pushes the solder slightly up the through-hole. The rise of solder into the plated through-holes is entirely due to the wetting forces set up by the capillary action of the solder in the through-hole.

The geometry of the through-hole, the component density, the number of layers in the board, and the temperature affect the rise of the solder. Generally, the higher the temperature, the greater the rise of the solder and thus; preheating becomes very important. Boards with sandwiched metallic layers may require top-side preheating.

After the solder has risen up the through-hole completely, the portion of the solder that extrudes from the top side of the printed circuit board is the first to cool and form a cap. The formation of a cap on the solder that rises into the through-hole plays a very important part in helping the solder stay in place in the through-hole while cooling down. Without support, the molten solder tends to fall back into the pot and, because of the simultaneous cooling due to the ambient air, forms icicles of solder hanging down from the joint sites.

The portion of the solder below the printed circuit board cools next and solidifies, while the center remains molten. The center of the joint may have vacuum holes (freezing vacuoles), but the quality of the joint is usually not sacrificed. Premature freezing of the bottom of the board is another undesirable phenomenon, because blowholes or pinholes form on top of the fillet or at the bottom of the board.

Zone 3 (point of exit and peel-back zone)

In third zone, the solder peels back from the board and falls back into the solder pot. Proper peel-back is essential to remove excess solder from the board. Surface and hydraulic forces affect the amount of peelback. Surface forces are either interfacial forces between the base metal and the solder, or cohesive forces within the liquid solder particles themselves. Hydraulic forces depend on the wave design, the impedance angle of the conveyor, the board land configuration, and the distribution of the thermal load on the board. The optimum design configuration of the board with the wave neutralizes the effect of hydraulic forces (i.e., the fillet is withdrawn at a static location, making the process similar to ordinary static dip soldering). The angle of the conveyor is kept at 3° to 7° to allow proper peel back.

3.4 Solder pots

The soldering operation takes place at the soldering station, which consists of the solder pot (Figure 3.13), the pump, and the nozzle. The solder pot is the container in which the solder is melted. The pot is made of cast iron, stainless steel, or occasionally, titanium, which can withstand a continuous operating temperature of 260° to 482°C and also bear both the load of the heavy molten solder and the corrosive action of the compounds in the solder. If temperatures will exceed 315°C, the pot is internally coated with oxide to protect the pot

Figure 3.13: Solder pots (*Electrovert*)

from the molten solder until the metal of the pot oxidizes itself and offers protection. Extreme care must be taken not to scratch the oxide layering off the walls while cleaning the solder pot.

The pot is usually heated by electric resistance heaters. Solder pots are often specially mounted on jacks to make them rigid and able to handle greater loads from the heavy solder waves. The size of the solder pot is usually much larger than actually needed for the soldering operation itself to help maintain a constant level of solder. Solder pots are also equipped with an automatic solder replenishment system for this purpose. Variations in the solder level in the pot cause joints either to become uncoated or to have excessive solder on them. Reduced solder levels result in dross contaminating the joints.

The main solder pot configurations are unidirectional and bidirectional; however, cascading and static solder pot configurations are also common.

3.4.1 Unidirectional solder pots

The unidirectional solder pot is designed to allow the solder to move in one direction only, relative to the board. Figure 3.14 shows a schematic of the wave and the various forces acting on the wave and the board. As the figure indicates, to prevent the formation of icicles of solder, the force $FR\cos m$ must be greater than $FR\sin m$. Otherwise, the tendency of the solder to wet and slip on itself increases, and the solder tends to fall back into the pot, leaving shallow fillets in the plated through-holes, especially in the case of long-leaded components. Icicles and solder shorts are the most common defects.

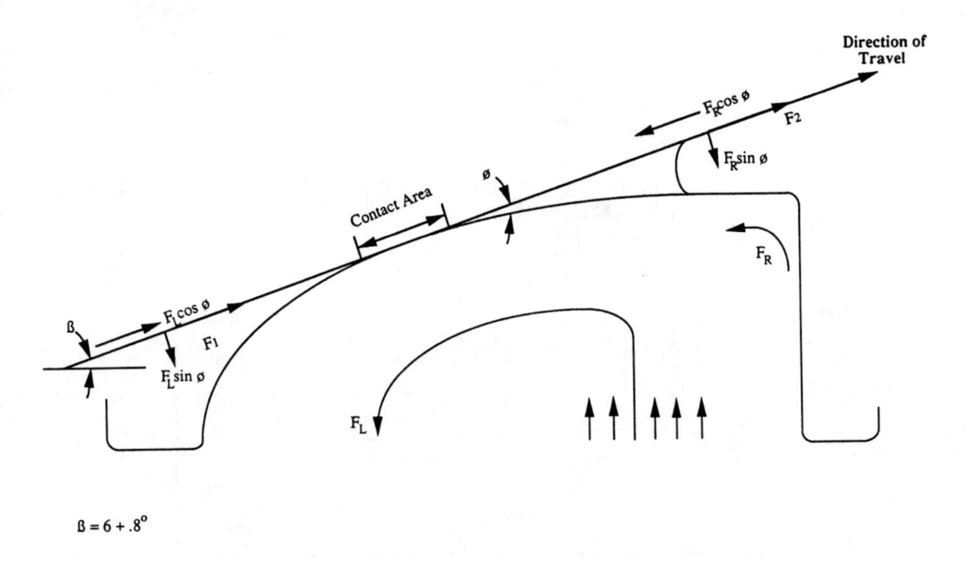

Figure 3.14: Wave form for a unidirectional solder pot [Lambert 1988]

3.4.2 Bidirectional solder pots

The bidirectional pot allows the flow of the solder wave both along the board travel and against it. The schematic of the wave formed in a bidirectional solder pot is depicted in Figure 3.15.

The part of the wave that moves against the direction of motion of the board wets the joints. The other part, which flows along with the board, helps pull away the excess solder from the board.

The forces encountered in this process are shown in Figure 3.15. The web of the wave must not get too far from the wave itself to avoid having the force $FY''\sin m$ exceed force $FY''\cos m$, causing icicles and solder shorts between the component leads. The conveyor is inclined to help separate the board from the wave as soon as it passes over the crest.

3.4.3 Cascading wave system

In the cascading wave system, a multiple waves are formed by flowing the solder down a convoluted surface. At each crest of the convolute, the solder wave rises to a peak. The printed circuit board also moves down the cascade.

Each time the wave forms a crest, it touches the board. The basic idea behind the cascading concept is to add solder to places where the run-off is high

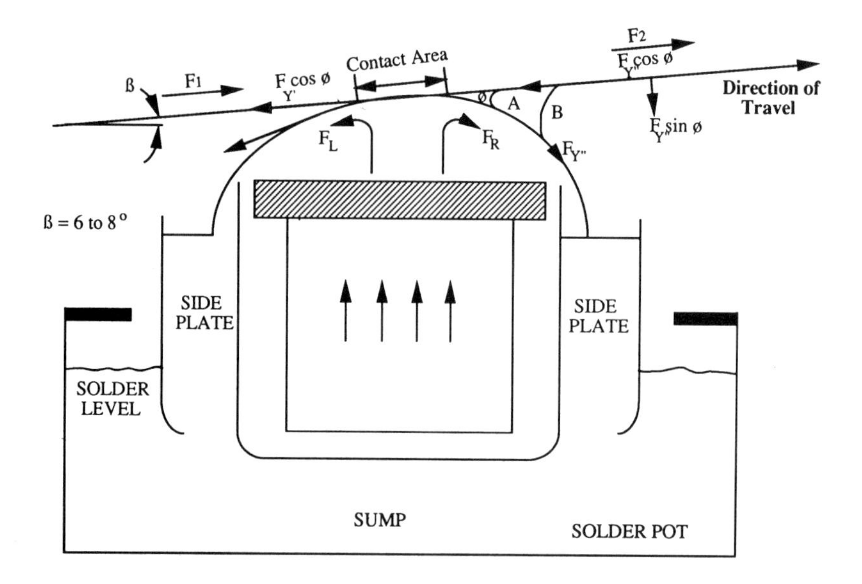

Figure 3.15: Wave form for a bidirectional solder pot [Lambert 1988]

and to remove excess solder. The cascading system is an interesting concept, but the idea never really caught on and few manufacturers provide this system.

3.4.4 Static solder pot

Compared to manual soldering of each component on the board, dip soldering is a major improvement, even though the sophistication and speed of the latest soldering technologies are more than a match for it. Still, static solder pots (Figure 3.13) are used in pre-tinning operations.

There are three main advantages of using static solder pots: (a) the amount of solder in the pot at one time is small, (b) there are no moving parts to maintain, and (c) the amount of dross generated in these pots is much less than in other types of solder pots. However, the static nature of the solder in the pot requires that dross generated during the soldering process, no matter how little be removed externally before the boards come into the pot. The conveyor has to be specially built to allow the boards to be brought into the pot at an angle. These disadvantages contribute towards the unpopularity of the static solder pot.

3.5 Pumps and nozzles

Waves are produced in the solder pot by means of a pump-and-nozzle arrangement. Special centrifugal pumps continuously pump solder through the nozzle into the solder pot, producing a solder wave. The height of the solder wave is controlled by regulating the motor speed or by the relative motion between the rotor and the static housing. Occasionally, both methods are used, mechanically moving the rotor for major adjustments and using an electric speed control for fine settings. A third alternative is to include a flap or valve inside the solder passage, and to vary the amount of solder in the nozzle by increasing or decreasing the opening of the valve.

The pump operates in a very hostile environment and, therefore, is ruggedly constructed. The high operating temperatures and constant exposure to molten solder make the use of bearings inside the solder pot virtually impossible; therefore, the bearings of all pump-and-nozzle systems are located well above the level of the solder in the pot. However, the extreme heat still does not allow ordinary bearings; special high-temperature bearings are required. The level of solder in the pot is maintained to protect the pump from chemical attack by the dross and to prevent the dross from sinking to the level of the pumping area and being sucked into the solder passageways.

Air motors are used by some manufacturers, but they do not have any special advantage over normal electric motors. Another technique uses an electromagnetic solder-pumping system. Here, the solder passes through a powerful electromagnet, inducing a current into the stream of solder. Due to an inductive reaction between the solder currents and the magnetic force of the electromagnet, the solder is propelled through small jet nozzles, producing a series of solder streams called a hollow wave. The main advantage of the electromagnetic system is the lack of moving parts; therefore, there is no need to lubricate any bearings, shafts, and so forth, and there is little problem with corrosion. Moreover, the high speed of the solder jets causes a scrubbing action on the joint, cleaning off contaminants more efficiently. These jets are also easier to control than jets produced by any other solder pumping system, and provide great flexibility in the types of boards that can be soldered. However, due to the low pumping forces available, large amounts of solder cannot be pumped at any one time.

3.6 Conveyors

The conveyor mechanism moves the boards through the fluxer, preheater, and solder wave. The pallet conveyor carries boards on flat pallets. The finger, or palletless, conveyor carries the printed circuit board through the various stages on blocks or long fingers that enable different sizes of boards to be transported. The following sections describe these two configurations in more detail.

3.6.1 Pallet conveyors

Many sizes of pallet can be used for soldering boards of different sizes simultaneously, thus reducing the cost of production. Additionally, the pallet also enables fixtures and cables to be incorporated in the same equipment assembly. However, only high-volume manufacturers can afford to incorporate a pallet conveyor system, which is economical only if the production run is reasonably long.

Pallets designed specifically for a particular job are rarely used in small-scale work. If flexibility is a must, most of the advantages of the pallet conveyor can be had by using simple tooling plates made of steel, tempered hardboard, or aluminum, routed to conform to the outline of the board to be soldered. These plates help protect parts of the board that must not come in contact with the solder wave.

The material of the pallet is an important consideration. If a reaction occurs between the pallet and the flux or solder at any time, the flux or solder becomes contaminated and must be cleaned, raising the cost of the process. Titanium or zirconium pallets are usually used, but aluminum, coated with an anodized or teflon coating, is used as a cheaper alternative. Steel, coated with an oxide layer to prevent the solder from attacking it, is typically used for heavy jobs requiring sturdier pallets. Because pallets must be absolutely flat for uniform soldering, they are tested every week for warpage, bending, or twisting. Large boards can be easily managed, using a pallet conveyor system, without causing any warpage in the boards. Tipping the pallet must be avoided because it causes the joints to skip the solder entirely, leaving unsoldered regions on the board. Vibration in the conveyor mechanism is another reason why some portions of the board may not be soldered. Therefore, the conveyor must be vibration-free.

Figure 3.16: Finger conveyors (*Sensbey, Inc.*)

3.6.2 Finger conveyors

In the finger-type conveyor Figure 3.16, hook-shaped fingers support the board and travel on chain drives into and out of the soldering equipment. The fingers catch hold of the boards as they are being fed into the equipment. After the soldering process, the fingers remove the boards from the machine and leave them on the next conveyor, which takes them to be cleaned. The finger must be flexible enough to accommodate temperature changes and board expansion, so as to not cause board warpage and solder floods.

Different board sizes are accommodated on the conveyor by adjusting the distance between the two rails of fingers. A hand-wheel or motorized mechanism is generally used for this purpose.

3.6.3 Conveyor drives

The conveyor drive generally has an adjustable speed of up to 8 to 10 m/s. Because speed consistency is important to maintain the accuracy of the conveyor over the complete operating cycle of the soldering machine, the conveyor drive is generally torque-controlled.

The drive is from a DC variable-speed or AC fixed-speed motor, with variation in speed caused by a mechanical drive. The speed control of DC drives is generally mounted at a distance from the motor. The controls are placed on the variable-speed motor.

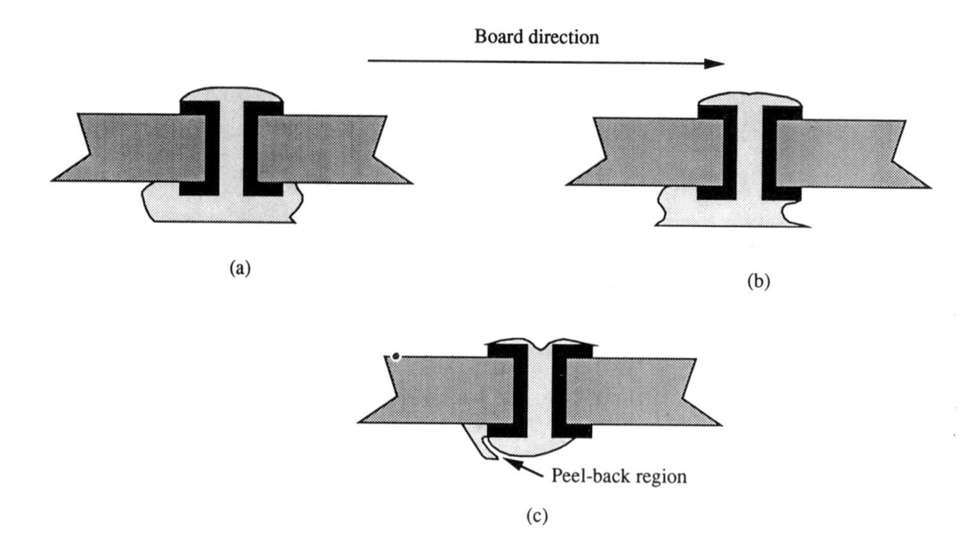

Figure 3.17: The effect of the solder wave on a leadless hole [Cummings 1992]

3.7 Future trends

3.7.1 Dynamic solder management

Dynamic solder management is a method of altering the flow around the soldering sites by the provision of barriers around the vias in the shape of U's or V's [Cummings 1992]. Without the barriers, the leads in the vias do not get thoroughly wetted.

Figure 3.17 shows how the molten solder, moving at a high velocity, first forms a bulbous cover above the via at the point of entrance. The solder then flows down along the board as the wave passes over the wetting region, and finally peels back into the reservoir. The solder left behind clings to the via walls and board bottom, forming a solder connection with a dimple in the top, due to the solder slipping down. If a lead is present in the via, the final shape of the solder fillet depends upon the position of the lead with respect to the via. As shown in Figure 3.18, a dimple forms on one side of the joint.

To avoid dimples, U- and V-shaped barriers are provided around the vias. This alters the flow of molten solder around the via and results in proper wetting of the leads. Figure 3.19 shows how the flow around the via changes due to open-ended U's and V's.

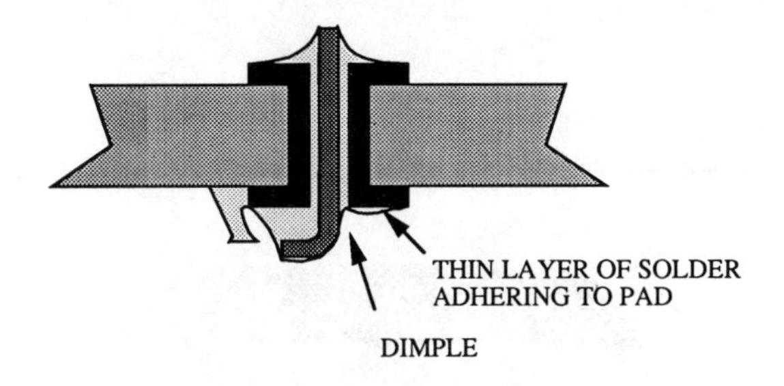

Figure 3.18: Solder fillet with lead

3.7.2 Ultrasonic soldering

Ultrasonic soldering is discussed with wave soldering because it uses a high-frequency blast of ultrasonic energy to form the solder joint. Ultrasonic soldering is the process whereby a molten solder alloy in contact with a compatible base material is ultrasonically agitated to generate an interface which results in metallurgical or mechanical bonding [Fuchs 1991]. Metallurgical bonding of the two metals occurs with the removal of dirt and oxides from the surface of the base metal by the ultrasonic wave. The oxides leave the surface of the base metal exposed to the solder, and ionic attraction between them causes a strong bond to form. In mechanical bonding, the ultrasonic explosion causes the base metal surface and the solder alloy to come into close proximity and thus form a bond.

3.7.3 Fine-pitch technology

Fine-pitch technology (FPT) is a step towards higher-density packaging with higher lead counts and finer lead spacings, generally sold under commercial names like PQFP (plastic quad flat-pack), TQFP (tabbed quad flat-pack), and VSOIC (very small-outline integrated circuit). In fine-pitch technology, unlike traditional soldering, the land size is almost the same as the lead because of the denser packaging. To ensure the reliability of the joint, as much solder as possible is applied, with an associated limit to prevent solder migration.

In some cases, lands-exposed boards with circuitry buried in the inner layers of the printed circuit board and exposed lands only on the outside layers are being used. The lands are connected to the sub-layers through small plated and filled vias inside the land area. The land areas can be made bigger, because

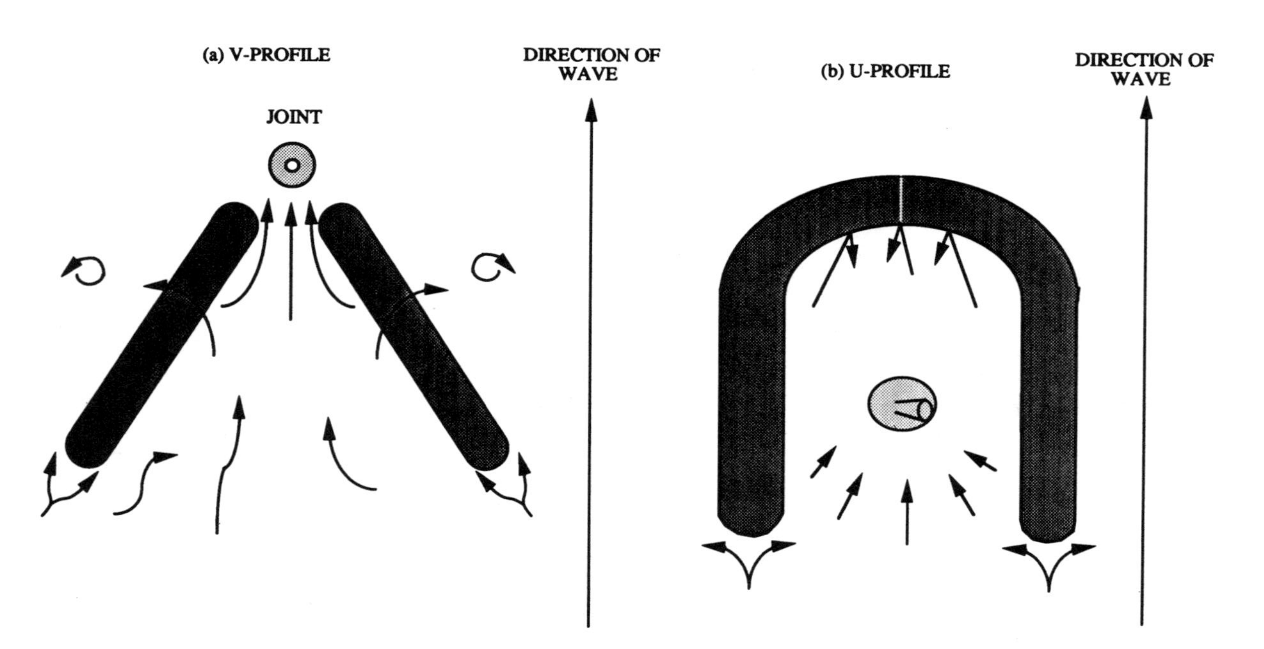

Figure 3.19: Change in solder flow direction due to the presence of (a) a V and (b) a U

none of the circuitry is in them. However, the added cost of via filling and the additional layers involved may make this approach uneconomical.

Fine-pitch technology has been the biggest challenge for wave soldering since its inception. In fact, the general belief earlier was that fine-pitch packages could not be wave-soldered. However, modifications in the fluxing, preheat, and solder-wave technology for fine-pitch applications, have enabled reliable wave soldering. Some of these modifications are described below.

Fluxers

To wave-solder fine-pitch boards, the flux material has to have a low solids content. Special stone and plastic aerators in foam-fluxing machines are used by some manufacturers of wave-soldering equipment to help increase the amount of foam in the flux, allowing greater flux-to-board contact. While some spray fluxers are being used, wave fluxers are probably the most promising because of their inherent ability to apply low-solids flux more evenly than any other fluxer. The disadvantage is that they leave a greater flux residue on the board. That must be cleaned later, and because manufacturers are now moving towards chlorofluorocarbon-free soldering (i.e., without any cleaning), this is a disadvantage.

Preheaters

The preheater must provide good temperature distribution and be able to repeat the same profile when required. The parameters generally considered are temperature rise and the length of the preheat stage. Among the various destructive effects of preheating is the breakage of packaged integrated circuit chips due to the expansion of trapped water vapor. For this reason, hot-platen preheating is generally not used for fine-pitch soldering; infrared and quartz lamp heating is more suitable. Another enhancement is the use of a sheathed preheater filled with magnesium oxide, with an element made of wound wire inside. Sheathed preheaters offer longer life and quality performance in terms of the preheating and the amount of maintenance required.

Solder waves

For fine-pitch components, two solder waves are generally preferred over one to ensure proper soldering of the joints. The waves can be generated in separate solder pots, thus allowing the use of different temperatures for each pot, providing more even temperature distribution over the board and ensuring reliable solder joints.

The first solder wave applies a large amount of solder to the board to ensure proper wetting and a full 360° application on the component lead profile. A vibratory or oscillatory nozzle can give the first wave the required agitation and turbulence to apply pressure on the board from all directions.

The second solder wave is generally provided by a solder chimney made of two-edge plates, separated by a distance through which the solder is allowed to flow. The flow of solder is horizontal both along and against its direction of travel. Additionally, the trailing edges of the two-edge plates rotate relative to the board to provide a greater breaking force and allow more effective peel-back of the excess solder on the board.

3.7.4 Controlled atmosphere wave soldering

Good soldering results require a minimum of solder bridges, icicles, and unsoldered joints. The demands on electronics technology are so great that any amount of rework results in a great loss of production potential and a consequent loss of the manufacturers market share. To avoid rework, manufacturers aim for zero-defect soldering, which demands good solderability of the joining metals, a controlled temperature profile over the board, an ideal solder wave, and no variation in environmental conditions at the time of soldering [Fodor 1990].

Flux is essential for proper wetting and for good soldering. Flux removes oxides from both the solder waves and the components to be soldered. If not removed, these oxides tend to promote the formation of solder bridges and icicles.

Fluxes leave residues that have to be cleaned later, using chlorofluorocarbon solvents that have been proven potentially harmful to the atmosphere. The Montreal Protocol of 1987 mandated reducing the use of chlorofluorocarbons in manufacturing processes gradually, and completely eliminating their use by the year 2000. In response, solder equipment manufacturers are going towards the development of "no-clean soldering processes" that do not require the use of fluxes. One method developed is the use of no-clean fluxes made out of substances that leave no residue after soldering [Manko 1990]. Another way of avoiding the use of flux is to perform the soldering process in an inert atmo-

Figure 3.20: A controlled atmosphere wave-soldering unit

sphere in which no oxygen is present to oxidize the molten solder; the use of flux then becomes unnecessary. The amount of oxides present on the components and board before the board comes in for soldering is small and can easily be removed by ultrasonic vibrations or by the flow pressure of the solder wave itself. (Figure 3.20 shows a controlled-atmosphere wave-soldering unit used in industry.) The no-clean process and other controlled atmosphere techniques are discussed in Chapter 5, "Cleaning and Contamination".

3.8 References

Capillo, Carmen. Surface Mount Technology, Materials, Processes and Equipment. New York: McGraw Hill Publishing Company (1990).

Cummings, Michael. Dynamic Solder Management Reduces Defects in Wave Soldering. *Electronic Packaging & Production* (July 1992), 74-78.

Fuchs, F. John. Ultrasonic Soldering in the Electronics Industry. Blackstone Ultrasonics, Inc. Jamestown, New York, NY (Internal publication, 1991).

Hwang, J. S. *Solder Paste in Electronics Packaging*, New York: Van Nostrand Reinhold (1989).

Lambert, Leo P. *Soldering for Electronic Assemblies* New York: Marcel Dekker, Inc. (1988).

Manko, Howard H. *Solders and Soldering.* New York: McGraw Hill Publishing Company (1992).

Manko, Howard H. Advantages and Pitfalls of "No Clean" Flux Systems. *Proceedings of the Technical Program*, NEPCON West (1990).

Woodgate, Ralph W. *The Handbook of Machine Soldering.* New York: John Wiley & Sons (1988).

Chapter 4

Reflow Soldering

Anupam Malhotra
CALCE Electronic Packaging Research Center
University of Maryland
College Park, MD

Wave soldering was the standard soldering process before surface-mount devices came into use. However, with the advent of surface-mount devices and increasingly finer pitches, manufacturers grew more concerned about controlling the wave-soldering process and sought better methods. Reflow soldering was found to be a good alternative.

The reflow soldering process involves remelting (reflowing) solder previously applied to a joint site in the form of a preform or paste, in order to create an attachment. No solder is added during the actual reflow. A better understanding of the reflow process can be attained by breaking it down into the following processes.

- **Solvent evaporation.** Solder paste contains solvents that regulate its consistency. In this phase of reflow, the temperature of the board is slowly raised in order to allow the solvent to evaporate. Typically, the heating rate is kept at 3°C/sec. The advantage of a slow heating rate is that solvent boiling and the formation of solder balls are avoided.

- **Chemical cleaning.** The flux in the solder paste is not activated. The ensuing chemical reactions help remove metal oxides from the surfaces of the metals to be joined. Some amount of contamination is also removed.

- **Melting.** Once the melting point of the solder paste is reached, the solder paste particles start melting individually. The liquid flux, formed in the second stage of reflow is replaced by the molten solder; due to its wetting action, a coat of solder is applied on the surfaces to be joined. The temperature must be maintained at the melting point to form a good bond. A higher temperature would cause the molten solder to move away from the desired joint location, along either the component lead or the printed circuit board, depending on whether convectional or radiational heating is used.

- **Fillet formation.** This stage is the most critical in the reflow process. The temperature must be regulated carefully to allow the melted solder particles to coalesce and then cool down. The cooling of the joint introduces surface tension effects in the solder volume, which allows the formation of a fillet around the component lead.

- **Cool-down.** Cool-down occurs either by conduction through the board layers, or by natural or forced convection to the ambient air. Forced convection is not generally used because it excessively stresses the boundary layers within the solder joint, reducing joint reliability. Strength considerations favor fast cooling, but the consequent stresses can be harmful. Therefore, the cooling rate is kept at 3°C/sec. — the same as the initial heating rate.

Heat can be applied to the solder preforms in a number of ways. The manner in which heat is applied to the printed circuit board differentiates various methods of reflow. Conduction heating was one of the earlier methods, but it soon gave way to convection (or vapor-phase heating). Radiation heating is the preferred method nowadays. Sometimes, a combination of radiation and convection heating is most effective. The goal is to move towards a process capable of providing a uniform thermal profile (see Figure 4.2) across the oven, subjecting the board to a smooth thermal gradient.

There have been significant developments in reflow soldering since the concept was introduced. Its usefulness in large-scale soldering of printed circuit boards has led to automation of most of the reflow soldering processes. Automation in solder-paste-dispensing technology involves the use of high-tech control systems to ensure repeatability and uniformity of the dispensed paste

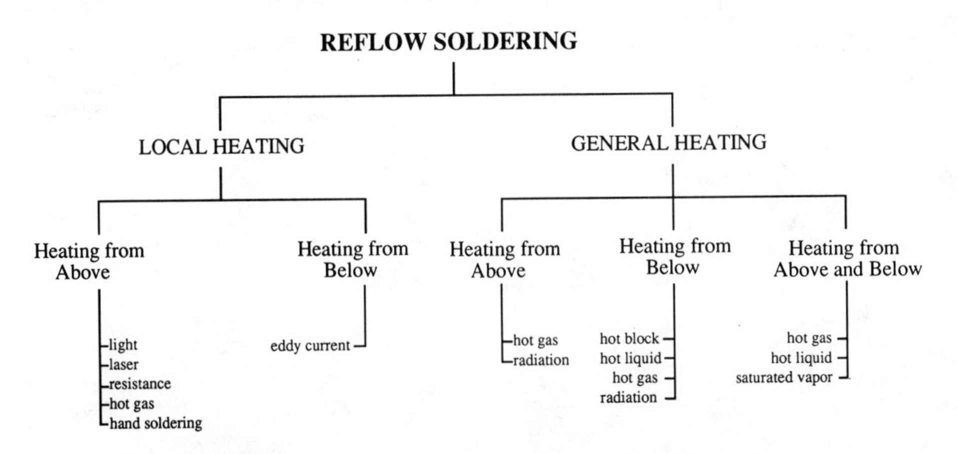

Figure 4.1: Classification of reflow soldering [Manko 1992]

and adhesives. These high-tech attachments include computer-vision systems, multiple-fluid dispensers, and computer controls. Some hand-held machines can work semi-automatically with electronic controls.

In a typical reflow-soldering process, solder and flux are added first, usually in the form of solder paste or cream. Components are then placed directly onto the wet paste, which acts as an adhesive. The entire assembly is then reflowed until the solder properly wets the surfaces to be joined, thus forming all the solder joints at one time. Additional steps, such as pre-tinning of the components, drying the solder paste, and cleaning flux residues after soldering are sometimes required. Pre-tinning chemically prepares the lead surface(s) for a metallurgical bond to the solder, and the flux removes the oxides formed while soldering. Reflow soldering also allows the attachment of surface-mount components to both sides of the circuit assembly, a process that can be accomplished in single or sequential reflow operations. (Figure 4.1 shows the various types of reflow-soldering systems being used in industry.)

The following sections describe the reflow soldering temperature profile, which must be optimized to get good soldering results; various methods of soldering; and reflow processes; classified according to the method of heat application — conduction, convection, radiation, or a combination.

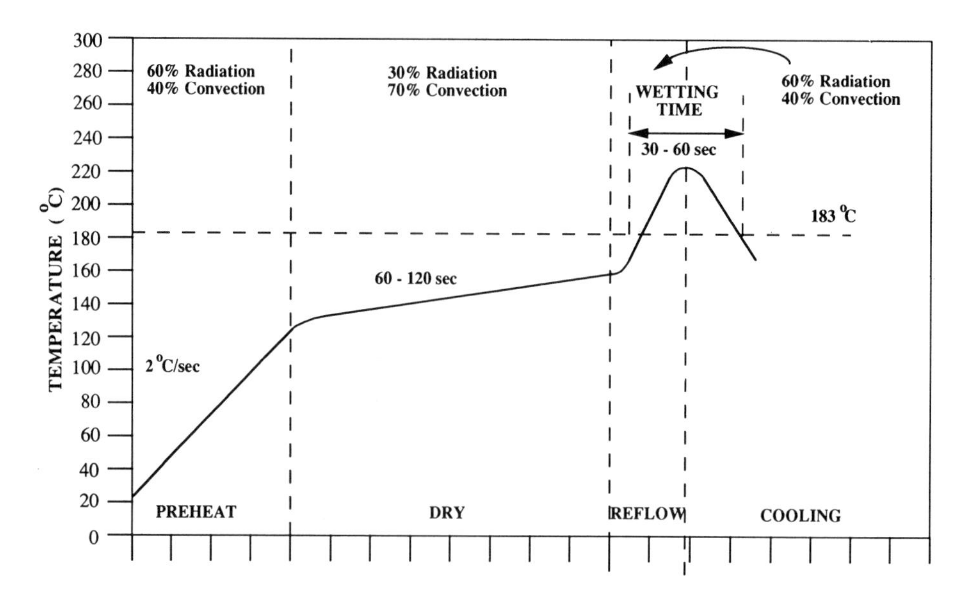

Figure 4.2: A typical reflow soldering profile [Research, Inc. 1992]

4.1 The reflow temperature profile

The temperature variation throughout the reflow soldering process must be controlled and optimized. (A typical reflow profile is shown in Figure 4.2). The goal is to ensure a high-quality, low-defect soldered joint.

The typical reflow profile can be divided into preheat, dryout, reflow, and cooling zones. Each zone has its own heating characteristics and rate. The preheat section heats the board from 100° to 150°C, with the temperature rising generally 2° to 4°C/sec. to minimize the thermal shock to the board and its components [Zarrow 1992]. Excessive thermal shock must be avoided to achieve reliable soldering; its effects include chip capacitor cracking and solder-paste spatter.

The dryout section is also called the soak or preflow zone. The main purpose of this stage is to fully dry the solder paste and activate the flux by heat before attempting reflow.

In the reflow zone, the solder paste actually melts and reflows to wet the joining surfaces completely and form a joint. Proper reflow requires the temperature to rise to 20°C above the melting point to form a good solder joint.

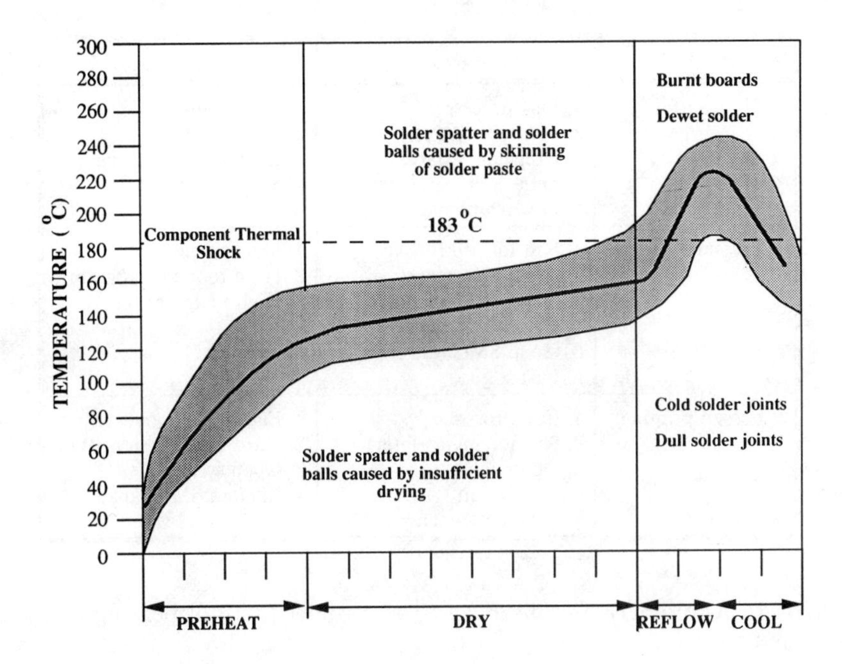

Figure 4.3: The reflow soldering profile band [Research, Inc. 1992]

The duration for which the solder remains above the melting point of the paste is called the wetting time (see Figure 4.2), and is usually 30 to 60 sec. for most pastes. If the normal wetting time is exceeded, excessive intermetallics may form and cause the joint to become brittle.

Figure 4.3 shows a typical reflow profile band for quality solder-joint production. The figure shows the band size as 25°C. However, the actual bandwidth depends on a number of factors, including the solder-paste material, the types of components, and the board material.

4.2 Solder-paste deposition

Prior to reflow soldering, solder paste must be deposited at the locations where the components will be attached to the circuit card. Adhesives may also be deposited to hold the components in place before reflowing.

Solder paste and adhesives are deposited by specialized machines ranging from elementary bottle-squeeze types to high-tech computer-controlled paste depositors incorporating machine vision to help locate solder-paste sites and determine the amount of paste and adhesive to be deposited.

Method	Advantages	Disadvantages
Pin transfer	1. Fast process 2. Soft tooling (machine) 3. Irregular surfaces 4. Some volume control 5. Low maintenance	1. Hard tooling 2. Open system 3. Limited SMC range 4. Slow set up
Pressure syringe	1. Soft tooling (machine) 2. Irregular surfaces 3. Good volume control 4. Closed system 5. Good SMC range 6. Fast set up	1. Slow process 2. Hard tooling (program) 3. High maintenance
Screen printing	1. Fast process 2. Soft tooling (printer) 3. Good volume control 4. Fast set up 5. Low maintenance	1. Flat surfaces only 2. Hard tooling (stencil) 3. Open system 4. Limited SMC range

Table 4.1: Advantages and disadvantages of solder-paste dispensing methods [Rowland 1990]

Solder paste is dispensed by three main processes: pressure dispensing, screen printing, and pin transfer. All of these processes can be accomplished either manually or automatically, depending on the scale of production and the amount of flexibility desired in the process. The dispensing process varies the amount of solder paste dispensed to hold different sizes of components in place. The amount dispensed depends on the component's height, type, and size; the composition of the paste; the board material; the space available on the board; and the type of equipment being used [Hodson 1992]. If the proper amount of paste and adhesive is not used, the component may fall off due to insufficient paste, or cause contamination or solder bridging problems due to excess paste. Pressure dispensing has an advantage over both screen printing and pin transfer, in that it uses a closed reservoir of solder paste and adhesive. Both screen printers and pin-transfer processes have open reservoirs, making them susceptible to contamination. Table 4.1 summarizes some of the advantages and disadvantages of these three dispensing processes.

4.2.1 Pressure dispensing

There are four main methods of pressure-dispensing solder paste and adhesives: the syringe method (generally used for dispensing solder paste dots and lines); the Archimedes rotary-screw method (also used for dispensing solder paste dots and lines); the peristaltic-action valve method; and the piston positive-displacement pump method (generally used for solder-paste dot dispensing). Each method is discussed here.

Syringe dispensers

Syringe dispensers are generally used to manually dispense solder paste and adhesives. A typical syringe dispenser (Figure 4.4) consists of a barrel filled with the solder paste and an air-compression or thumb-operated plunger. The plunger moves in the barrel to squeeze out the paste through a blunt-tip needle. The barrel in most syringes is not tapered. Instead, precision-molded rubber stoppers are fit in the top of the barrel to allow a uniform distribution of the paste through the tip.

The advantages offered by syringe dispensers include the following:

Figure 4.4: Automated syringe dispenser (*Westinghouse Electric Corp.*)

- Solder paste can be dispensed in hard-to-reach places and on odd-shaped or irregular surfaces.
- Solder paste can be directly applied to surface-mount components after the through-hole components have been soldered in place.
- The method is flexible and easily adapted to new designs, because the person doing the dispensing needs to know only what new components need paste.
- Syringes do not have to be cleaned because they are usually disposable.

Disadvantages of syringe dispensers are that manual methods are slow and the operator has no control over the amount of paste dispensed — the repeatability of the process and thus, the quality, depends on the operator. These disadvantages can be remedied, in part, with a time-pressure dispenser that uses a pneumatic pump to apply consistent pressure to the solder paste and adhesive mixture with a timed air-pressure signal. Thus, the dispensing process is faster, as well as neater and more uniform.

Archimedes rotary-screw dispensers

The Archimedes rotary screw-type dispenser uses pneumatic pressure to push the solder paste into the barrel of the dispenser. Then a screw rotates to control the amount of adhesive or solder paste pushed into the needle. This feature makes it possible to use the Archimedean screw to dispense different dot sizes. The process is generally computerized.

Peristaltic action valve dispensers

Peristaltic dispensers use pinch-tube displacement pumps that dispense the solder paste or adhesive from a cartridge. The pinch tube controls the amount of pressure applied on the cartridge and the amount of paste dispensed. The pressure difference does not affect the shape of the dispensed paste or its quality.

Piston positive-displacement pump method

In the piston positive-displacement pump method (shown in Figure 4.5), solder paste is dispensed by an air-pressure pump. The speed of the process is high, and solder paste or adhesive dots can be quickly applied on the board. Because the approach is unstable when dispensing a line of solder paste, a series of overlapping dots of paste are applied on the board to approximate a continuous line. Such a line is nonuniform, which can cause problems during reflow.

Automation in solder-paste-dispensing technology involves the use of high-tech control systems to ensure repeatability and uniformity in the dispensed paste and adhesives. These high-tech attachments include computer-vision systems, multiple-fluid dispensers, and computer controls. Some hand-held machines can work semi-automatically, using electronic controls.

4.2.2 Screen printing

The process of solder paste and adhesive screen printing is similar to the screen printing process used by printers and publishers. The finished board is placed on a holder and a stencil (or screen) is placed over the board. The stencil has all the non-solder paste points covered with an emulsion paste. A squeegee blade is passed over the screen (Figure 4.6), forcing solder paste and adhesive through the screen. The paste is deposited on the board at the points where there is a gap in the emulsion. If different thicknesses of paste are to be deposited, more than one screen must be used. The three main parts of a screen printer are the substrate holder, the screen or stencil holder, the stencil, and the squeegee.

The circuit-board holder

The circuit-board holder must hold the printed circuit board during the screen printing process to prevent any shift in its position and deposition of the solder paste in the wrong places. Additionally, the substrate holder has to locate the board with respect to the screen and act like a screen-printing jig. Vacuum cups or mechanical clamps hold the board in place. The use of vacuum cups requires the flow of air below the board to be directed so as not to affect the solder paste being deposited.

Double-sided boards are more difficult to hold because they require clamping at the edges. To reduce bending or bowing, additional supports must be provided to keep the board level.

The stencil holder

The stencil holder holds the stencil in place during the entire screen-printing operation with the help of clam shells, or vertical parts that open to allow

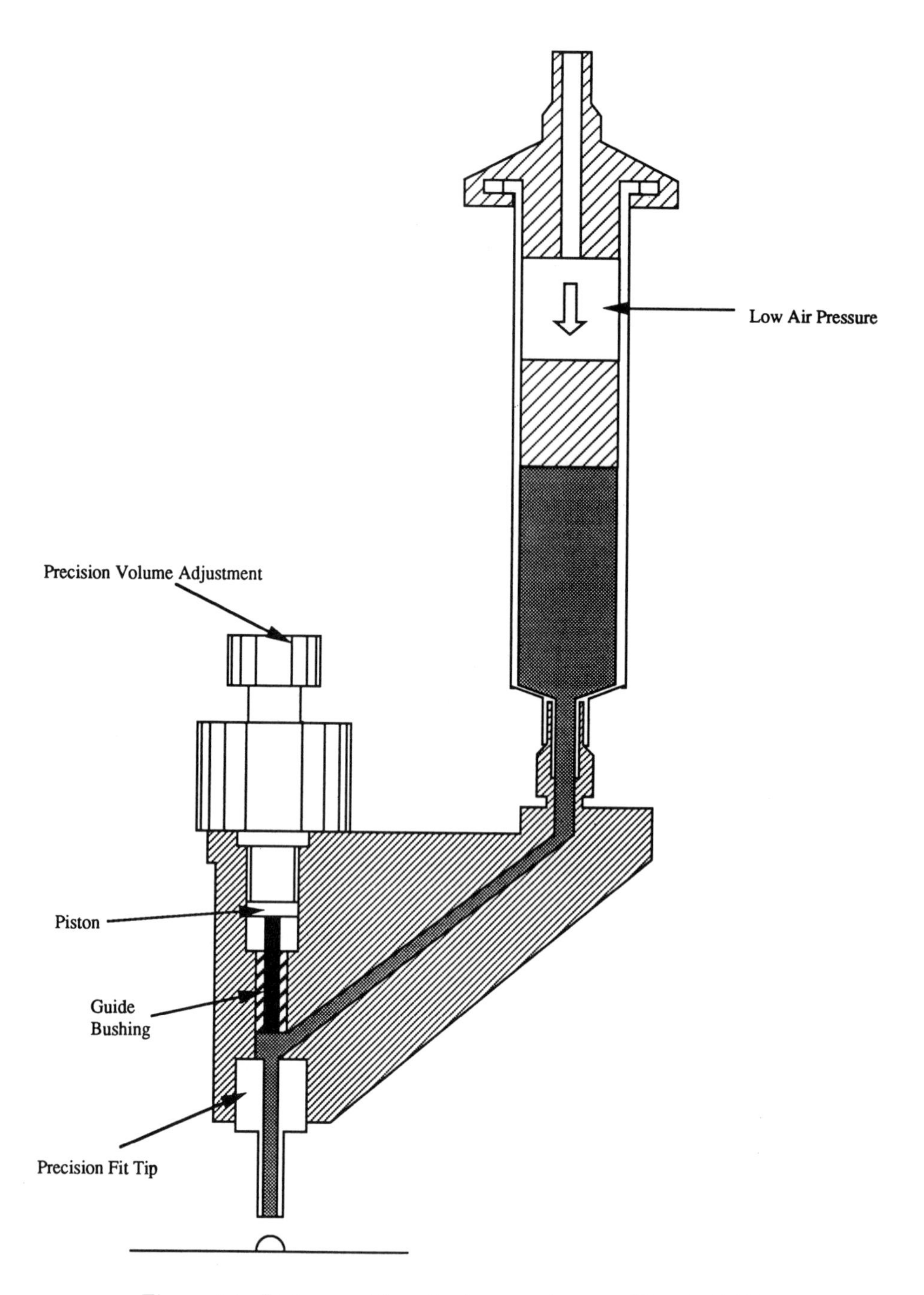

Figure 4.5: Piston-positive displacement pump [Engel, 1991]

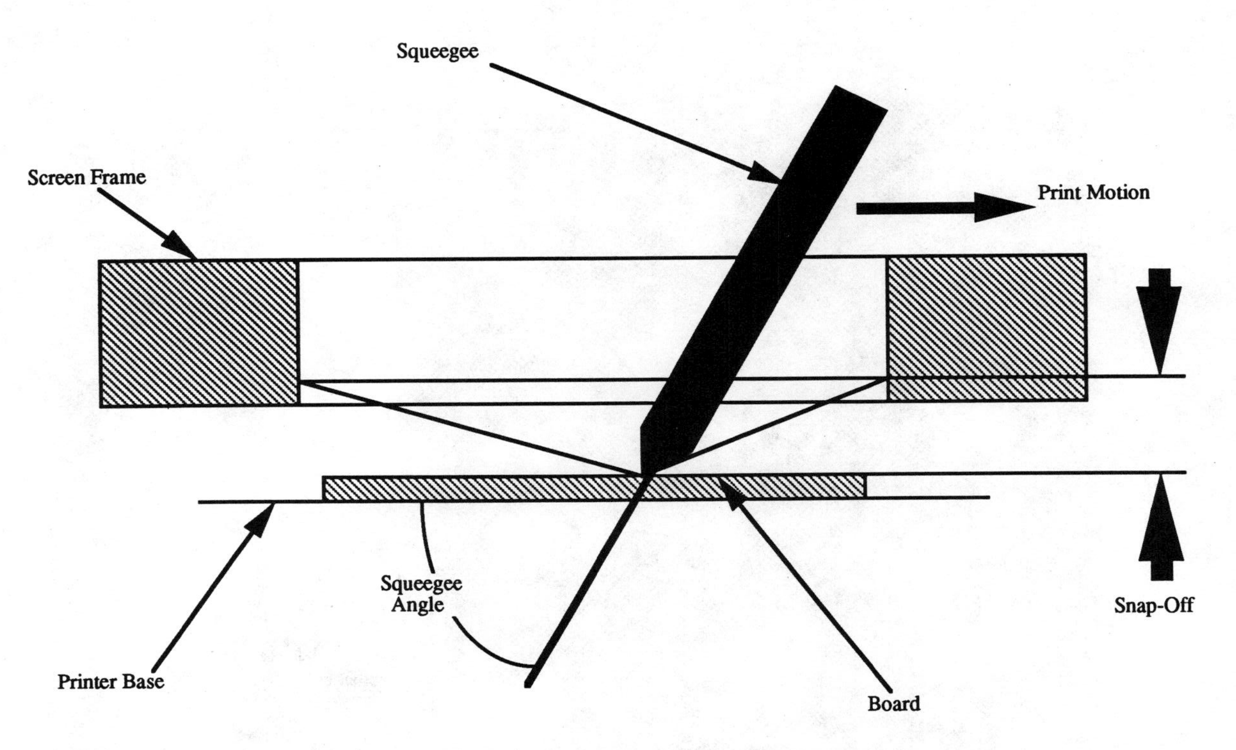

Figure 4.6: Schematic of the screen printing process

Figure 4.7: Screen printing solder paste on printed circuit boards (*Westinghouse Electric Corp.*)

Figure 4.8: Vision alignment for a screen printer (*Westinghouse Electric Corp.*)

adding or removing the screen. The registration of the screen to the board is important, because even a small deviation from the proper position causes all the solder paste points to be misplaced. Clam shells, although relatively inexpensive, do not provide good registration accuracy. A better alternative is to use vertical posts on which the holder travels up and down to add or remove the screen — a much more accurate registration method.

A screen cannot be placed directly on a board because the solder paste print on the board may become smudged. Instead, the board is placed at a slight distance from the screen, called the snap-off. When the squeegee passes across the board, the screen is pressed into contact with the board and it lifts back up after the pass. In this case, the thickness of paste deposited can be greater than the thickness of the stencil [Enterkin 1991]. Typically, the snap-off ranges from 20 to 50 mil for screens, and slightly less for stencils.

The screen or stencil

The stencil or screen is generally made of a foil of brass or stainless steel held to a polyester or stainless-steel mesh. The assembly is then attached to a cast aluminum or tubular frame after the stencil is stretched out. The polyester border helps to ease the stresses on the stencil when pressed down by the squeegee, thereby increasing the life of the stencil.

The amount of solder paste deposited on the board depends on the ratio of the stencil wall area to the paste volume. To use stencil printing for fine-pitch solder-paste deposition, which has a large wall area-to-paste volume ratio, the stencil has to be stepped in design with the top surface etched downwards to reduce its thickness. In the step-down stencil, all solder-paste patterns greater than 50 mils are etched in the foil's original thickness and smaller land patterns are etched in the foil's thinner, stepped-down locations. These land patterns for fine-pitch components are made so as to leave a border of reduced thickness around the pattern allowing the squeegee blade to properly deflect into the cavity and deposit the solder paste. Another method is to design a multilayer stencil by making fine holes in the lower layers and progressively wider ones in the layers above to better control the amount of solder paste deposited.

A stencil design method that is proving successful is staggering the stencil (i.e., developing a stencil in which the shape of the openings is altered). This prevents the flattening of the wet paste and subsequent bridging of the solder joints due to reflow, which occurs when the components are placed on the board after dispensing. The staggering is in the shape of triangular or tear-drop-shaped holes, each of which is inverted with respect to its neighbor. Solder bridging is prevented by depositing the solder paste so that the wide portion of one dot is adjacent to the narrow part of the next one, reducing the chances of their joining. While the problem can also be eliminated by using a solder mask, this involves additional costs.

The squeegee

The squeegee blade is designed to allow proper solder-paste and adhesive deposition on the board. Fine-pitch applications require that the blade be made of a hard material (up to 100 durometer) to allow all the solder paste to be deposited on the board. Generally, polyurethane plastic is used for these blades. The important parameters considered in designing a squeegee are the pressure, blade level, attack angle of the blade, speed of movement of the blade, and stroke length. If the squeegee blade presses the board too hard, the deposited solder paste comes out of the holes in the screen by a process called bleeding, or is scooped up by the blade and is not completely deposited. Moreover, a high amount of friction between the blade and the screen causes the board to be displaced, resulting in the paste being deposited in the wrong places. These factors call for the use of metal squeegees impregnated with a lubricant to reduce friction. Metal blades have greater wear resistance and can maintain a greater degree of flatness on the screen. The levelling produced by the blade prevents any paste from remaining on the screen rather than being deposited on the board.

The blade is either diamond-shaped or triangular, with a trailing edge. The diamond shape is generally used for small and flat boards; triangular squeegees are mostly used in the case of large and irregular boards. The attack angle (the angle between the normal to the surface of the screen and the squeegee blade) is about 45° to 60°.

The speed of movement of the blade is important, because slow speeds cause the solder paste to smudge, and fast speeds do not provide enough time for the paste to squeeze into the gaps in the screen and get deposited on the board. The general speed used in fine-pitch applications ranges from 1 to 2 in./sec., and in coarse-pitch applications from 2 to 6 in./sec., The stroke length refers to the distance travelled by the squeegee, and depends on the size of the board and the type of screen used.

4.2.3 Pin transfer

The pin transfer process of solder paste application involves dipping pins in a reservoir of solder paste or adhesive and applying their ends to the component sites on the board. The amount of paste deposited can be controlled by the pin size, the shape of its tip, the viscosity of the paste, and the distance between the pin and the board when the paste is applied to the board.

Pin transfer is very useful in high-density applications, because a whole array of pins can be simultaneously dipped into the paste and placed on the board. Manual pin transfer is generally used when prototypes of boards are being made, to economize on the process development cost. The greatest disadvantage of this process is the possibility of contaminating the reservoir by repeatedly dipping pins in it.

4.3 Conduction-based reflow soldering

4.3.1 Conductive (hot-bar) reflow soldering

The hot-bar reflow soldering technique requires a heater bar to directly heat the leads on a printed circuit board (Figure 4.9). The heater bar operates in much the same way as the manual soldering iron, except that the hot heater bar is not wetted by the solder. Therefore, freshly soldered joints are not damaged when the heater bar is withdrawn. The heater bar is capable of soldering multiple leads to the printed circuit board simultaneously. Hence, the hot-bar process has a very wide application in places where wave soldering is not possible and the use of other reflow soldering techniques is not justified.

The heater-bar soldering operation has three main operation variables: the heater-bar temperature, the dwell time at heat, and the clamping load. The

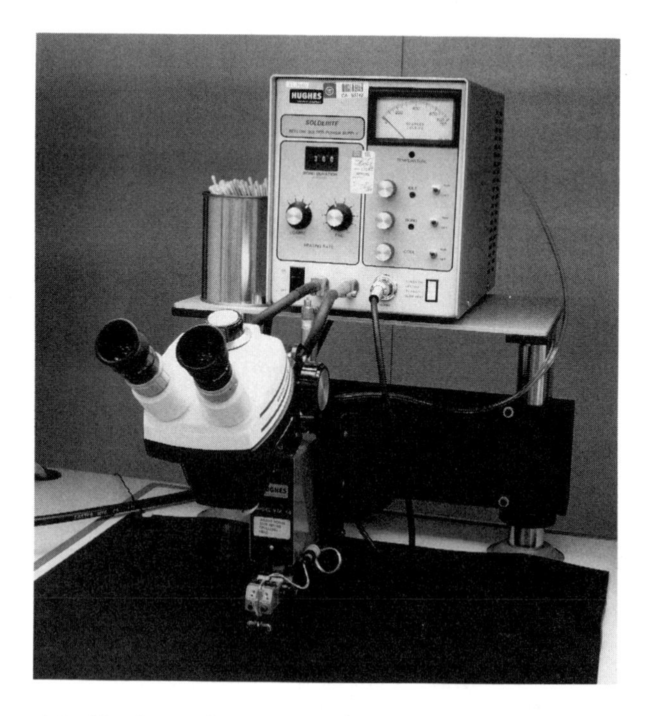

Figure 4.9: Hot bar reflow station (*Westinghouse Electric Corp.*)

lower these three parameters, the less the chances of damaging the printed circuit board during the process. Higher values of these parameters result in faster joint formation, but may also result in increased solder defects.

4.3.2 Conductive-belt reflow soldering

The conductive-belt reflow technique is used for low-volume, small, single-sided printed circuit boards. The boards are transported on very fine-meshed, flat-headed belts to the soldering station. These belts and circuit boards are heated by passing over resistance-heated platens. The temperature profile required for effective solder is achieved by passing the printed circuit board over a series of independently controlled heating and cooling platens.

4.4 Convection-based reflow soldering

4.4.1 Vapor-phase reflow soldering

The vapor-phase reflow-soldering process, introduced in the mid-1970s, was initially used primarily for the mass reflow of wire-wrappable pins into backplane assemblies, connectors to multi-layer boards, and component and lead frame attachment to hybrid circuits.

The product to be reflowed by the vapor-phase process must have solder pre-placed on the board as preforms, prior-plating, or solder paste. The product is then immersed in the saturated vapor above a boiling fluid. For the soldering process to occur effectively, the mating parts must be above the melting temperature of the solder to wet the joints. (Figure 4.10 illustrates the process.)

The greatest advantage of the vapor-soldering process is its temperature control over the entire board and rapid heat transfer from vapor to board and other components. The circuit boards are usually preheated to about 100°C to protect the board and components from thermal shock on entry into the vapor chamber. After preheating, the boards are taken into the vapor chamber. Freon vapor was very popular when vapor-phase soldering was developed, but fluorinated pentapolyoxypropylene (E5) and perfluorotrianylamine (FC-70) are now more common. These vapors condense on the board, components, and joints, releasing the latent heat of vaporization. With the release of the latent heat, the temperature increases on the entire board and on the components.

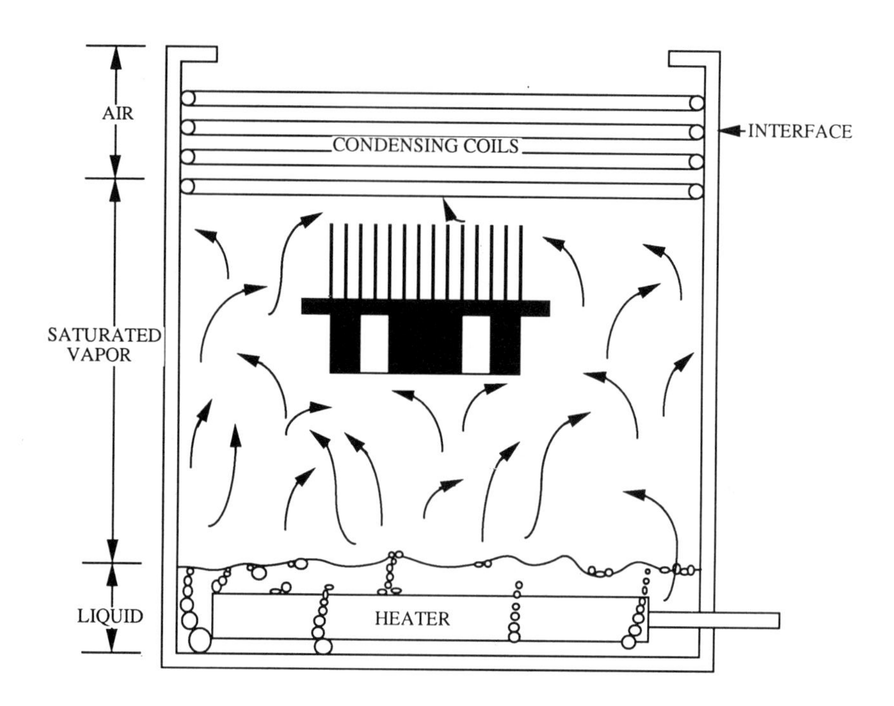

Figure 4.10: Vapor-phase reflow-soldering equipment

The board is then brought out of the vapor chamber and allowed to cool. During this time, the solder solidifies, forming all the joints on the board. Rapid cooling of the solder joints is advised so as to avoid the formation of large grains in the solder which ultimately weaken the joint.

There are many advantages to the use of the vapor phase process for soldering. The temperature in the chamber is always constant and requires no independent control, because the saturated vapor above a boiling liquid has the same temperature as the boiling point of the liquid. The fluids used have a boiling point of 215°C (420°F) and, therefore, the product is protected from overexposure to excessive temperatures. The temperature in the chamber is considerably lower than the temperature of many other processes, as the energy transfer is not due to the large temperature differences but to the phase transformation of the vapors.

Oxidation of product or flux is practically eliminated since the board is in an atmosphere of saturated, contamination-free vapor. Moreover, air, the main cause of oxidation, is eliminated due to the saturation level of the vapors. Thus, small levels of mildly activated fluxes can be used and can be easily cleaned from the product; these fluxes undergo neither oxidation nor polymerization.

The process is very effective for soldering boards with components in an unusual configuration, as the vapors condense evenly on all the surfaces on the board. Modern soldering systems incorporating vapor-phase technology offer single-in-line as well as batch-production systems that do not use chlorofluorocarbons. In-line systems include a preheater section, which is optional in batch-type systems. The latest machines have conveyors that allow horizontal movement of the boards, eliminating the possibility of board and component slippage. The quick cool-down of the joints gives the solder a fine-grain structure, resulting in the greatest joint strength. Perhaps the greatest advantage offered by today's vapor-phase reflow equipment is the incorporation of vapor recovery systems that save enough to recoup the equipment cost in about nine months or less.

Vapor-phase technology is used in 20-mil-pitch density boards without any problems, and in some cases these techniques have produced good results in finer-pitch densities, too [Paavola 1990]. The greatest advantage offered by the vapor-phase process is the control over temperature. The vapor's temperature cannot be hotter than the liquid's boiling point; therefore, when the two are in equilibrium, there is no inherent danger of overheating and burning the board and its components. In contrast, neither IR nor laser-reflow techniques can guarantee such control over the temperature profile on the board. Vapor-phase technology also has the advantage of better and faster board curing, with only a few minutes or seconds required for complete curing, compared to the many hours required in the case of convection ovens.

4.4.2 Condensation soldering

In condensation soldering, an extension of vapor-phase soldering, condensation coils help prevent the boiling vapors from the solder surface from escaping. These condensation coils are typically water-cooled and condense the hot vapors rising to the top (Figure 4.11). Thus, there is always a continuously replenished saturated vapor zone.

Mist formation is still a cause of vapor loss. The condensation of the vapor into fine droplets on meeting the cold air layer sometimes causes a region of mist above the vapor region. The convective flow causes the misty vapor to be lost into the surroundings. The solution to this problem is the use of the vapor blanket technique, in which a less dense, lower-boiling-point, inexpensive secondary vapor is floated upon the primary saturated vapor. Trichlorotrifluoroethane vapor, which is inexpensive and also chemically and thermally inert and stable with respect to both the primary vapor and the workpiece, is commonly used as a secondary fluid.

Figure 4.12 shows a condensation soldering system. The primary liquid is boiled continuously from the sump; the liquid rises up to the primary coils, only to condense and fall back into the sump. There is a similar cycle even in the secondary fluid region. The secondary fluid, which is less dense than the primary, is heated by the hot primary fluid and rises to meet the secondary condensing cooling coils. The vapor then condenses and recirculates in a region above the primary fluid. The secondary vapor is lost to the atmosphere and has to be continually replenished from a reservoir. Due to the significant temperature and density gradient between the primary-secondary and secondary-air layers, their interfaces are well defined and convective mass flow between the layers is minimal.

Mixing and diffusion of the secondary and primary vapors causes the temperature of the vapor blanket to actually be much higher (190°F) than the boiling point of the secondary fluid (117.6°F for trichlorotrifluoroethane). Thermodynamics shows that the higher the temperature, the higher the weight percent of the primary fluid in the vapor blanket. This process is cumulative, and temperatures far higher than required for reflow can occur. Therefore, operating at temperatures as low as possible is desirable.

Condensation soldering typically requires a vapor tank with primary and secondary vapor zones, a heating arrangement for the vapors, condensation coils, a filtration system, a programmed conveyor system, and an exhaust system. (A schematic of a typical condensation-soldering facility is shown in Figure 4.12.) The various parts of the vapor-phase reflow soldering equipment are described below.

- **Vapor tank.** The vapor tank, a welded construction that is usually free-standing without external support, contains the entire vapor system. The dimensions of the tank depend on the size of the part to be soldered.

  Printed circuit boards, properly pre-treated, fluxed, and loaded onto the carrier, enter the tank from the top. On entering the vapor tank, the boards pass through the vapor blanket and stop in the primary vapor

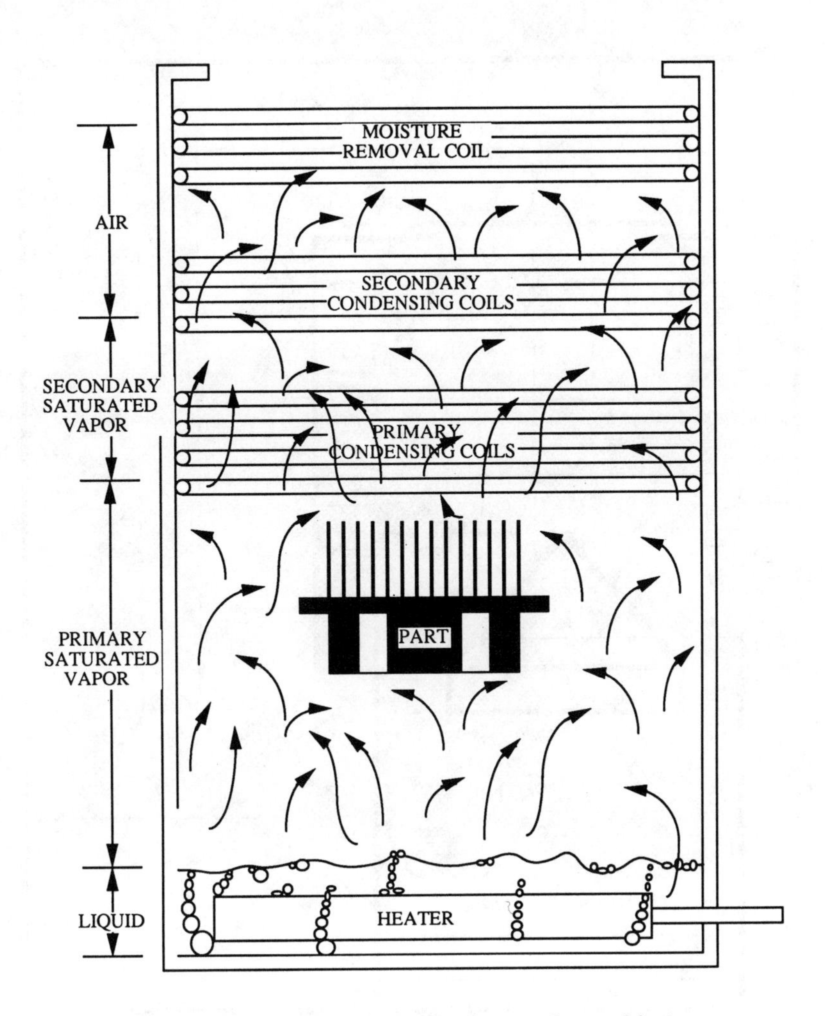

Figure 4.11: Secondary vapor-phase reflow soldering

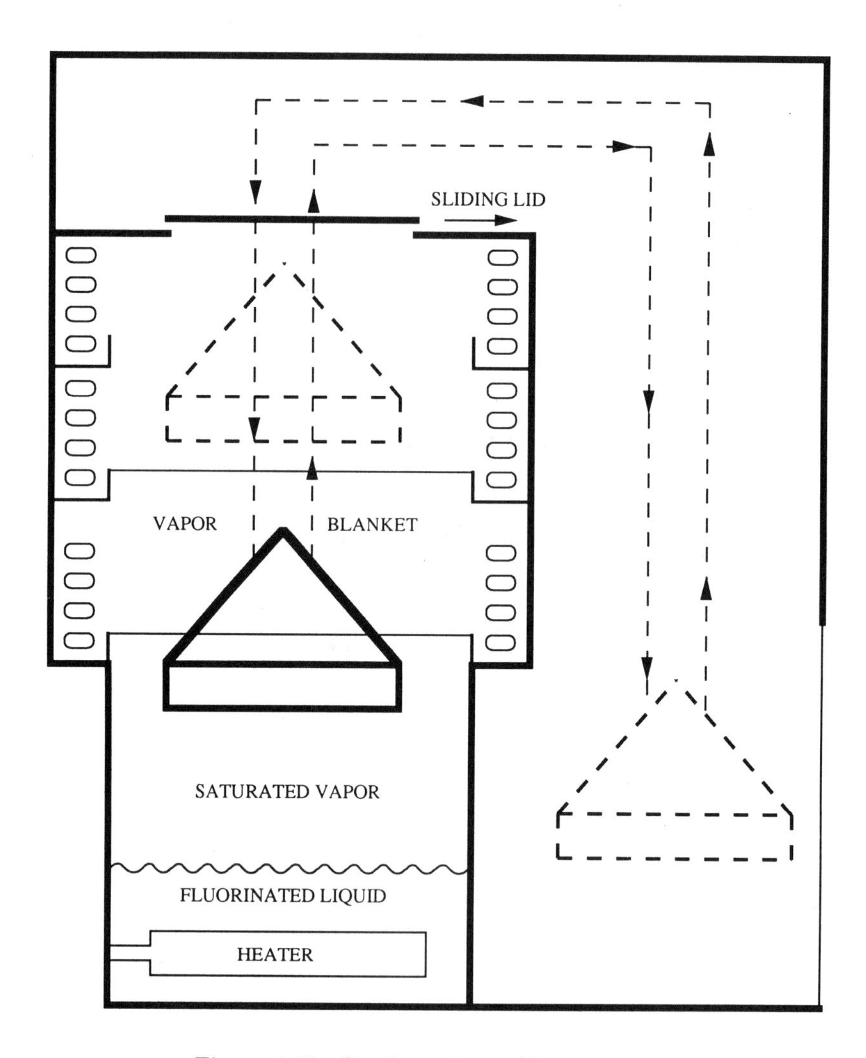

Figure 4.12: Condensation-soldering system

for a dwell, typically for a minute or less. The carriage is then raised to the vapor-blanket region and held for about 1.5 min. to allow any remaining film of expensive primary fluid to boil and drain back to the primary sump. The carriage is then brought out of the tank, along with the soldered part.

The clearance between the printed circuit boards and carriage and the thickness of the vapor regions themselves determine the vertical dimension of the tank. Practical clearances have to be provided between (a) the printed circuit boards and the boiling surface, (b) the printed circuit boards and the vapor interface when in the dwell position in the primary vapor, and (c) the lower surface of the part and the vapor interface when in the vapor blanket.

The vapor zones are made deeper than recommended to allow different sizes of printed circuit boards to be reflowed. However, this arrangement increases the cycle time, due to increased time for travel.

The vapor tank usually has a flat-bottomed collecting trough for the condensed primary fluid. The trough acts as a reservoir for the hot primary condensate, which is used as an energy source for evaporating the secondary fluid to maintain the vapor blanket. When such an arrangement is not present in the vapor tank, the energy for vaporizing the secondary fluid is provided by the tank wall of the primary fluid region or the primary fluid itself, as the condensed secondary fluid trickles down the side of the tank. The tank wall usually has a water jacket at the primary-secondary interface to prevent the secondary fluid from being heated by conduction from the primary section of the tank. The lower region of the tank wall is insulated with mineral or glass wool.

- **Heaters.** The energy required for the generation of saturated vapors is provided by low-watt electric immersion heaters, which avoid thermal degradation. The heaters are either continuous, or a combination of continuous and reserve banks. The continuous power bank maintains two stable vapor zones during idling, and the reserve power bank regulates temperatures during operation. These reserve power heaters are controlled by a thermostat located just below the surface of the vapor level. The heaters are switched on and off, based on the temperature in that zone which may vary when large parts enter the tank, the temperature of the vapor drops and extra heat is required to maintain the temperature at a constant level. Other controls are also provided to the heater circuit to prevent overheating. This is especially important when there is a drop in the vapor levels.

- **Condensing coils.** Condensing coils are helical, stainless-steel coils lining the inside wall of the vapor tanks. These coils are present only at levels where the vapors condense. The vapor-zone level during idling (when the condenser is not in use) is kept at least at the level of the lowest turn of the coil for that vapor region.

  Water from the water chiller or plant cooler is circulated through the secondary and freeboard moisture-condensing coils at 4o°-60°C. The temperature of the water is kept above the dew point to avoid water vapor condensation. The coils in the secondary vapor region are coated with fluorocarbon to prevent corrosion. The primary condensing coils circulate water at 50°C or above to avoid condensing the secondary vapors (the boiling point of the secondary fluid is generally 45°-48°C). The water requirements for the primary coils vary. Water is supplied by a surge tank electrically heated to keep it at constant temperature. Heating the surge tank is needed only at start-up; once the vapors are established, cooling is required to remove the heat of condensation. The flow of the tempered water is from the surge tank to the water jacket, then through the primary condensing coils, and finally back to the surge tank through a heat exchanger that keeps the water in the surge tank at 50°C. The flow to the heat exchanger is controlled by a solenoid valve triggered by a temperature sensor at the outlet of the surge tank or the inlet of the primary coil.

- **Secondary-injection system.** The level of the vapor in the secondary vapor blanket is controlled by a continuous recirculating injection system. The fluid is injected by means of injection ports between the condensing coils and the middle of the secondary vapor blanket in measured quantities. The injection ports are symmetrically arranged to minimize disturbances due to rapid vaporization as the fluid flows down from the injection ports toward the primary fluid. Flowmeters and needle valves regulate the rate of flow.

The vapor is recirculated by collecting the condensed vapors, separating the fluid from any moisture, and sending the separated fluid back to the secondary vapor tank. Because some secondary fluids are corrosive, chemical filters must be used in the circuit to neutralize the corrosive effect.

- **Filtration system.** The printed circuit boards to be soldered carry rosin flux and other foreign material when they enter the vapor tank. A large part of this material is washed off during the soldering process and enters the vapor circuit. Because these materials can cause problems in the circuitry on the board, they must be removed. Continuous filtration, distillation, bath filtration, or chemical treatment of the vapor accomplishes this. The most widely used method is batch filtration, which involves cooling the vapors to precipitate the rosin, followed by filtration to remove it. This operation is performed only when the equipment is shut down. (Figure 4.13 shows the continuous and batch rosin-filtration circuit.)

- **Conveyor system.** The choice of the conveyor system depends on the printed circuit board shape and size. The important features to be accommodated are low conveyor speeds (10 ft./min.) and the ability to operate without lubrication, as the secondary fluids commonly used act as degreasing agents. The tanks allow entry into the conveyor through a sliding roof or, in the case of hooded tanks, the absence of the roof. Other features are an efficient exhaust system for fumes from processed printed circuit boards, and the electrical and operating controls for the system.

4.5 Radiation-based reflow soldering

Radiation techniques reflow the solder with radiation heat transfer from a heat source. Heat is transferred directly by electromagnetic waves to the object to be heated without heating the intervening medium. Among the factors affecting the rate of radiation heat transfer are the emissivity of the source of heat, the absorptivity of the target, the temperatures of the source and target, and the view factor (the fraction of emitted energy that reaches the targeted object).

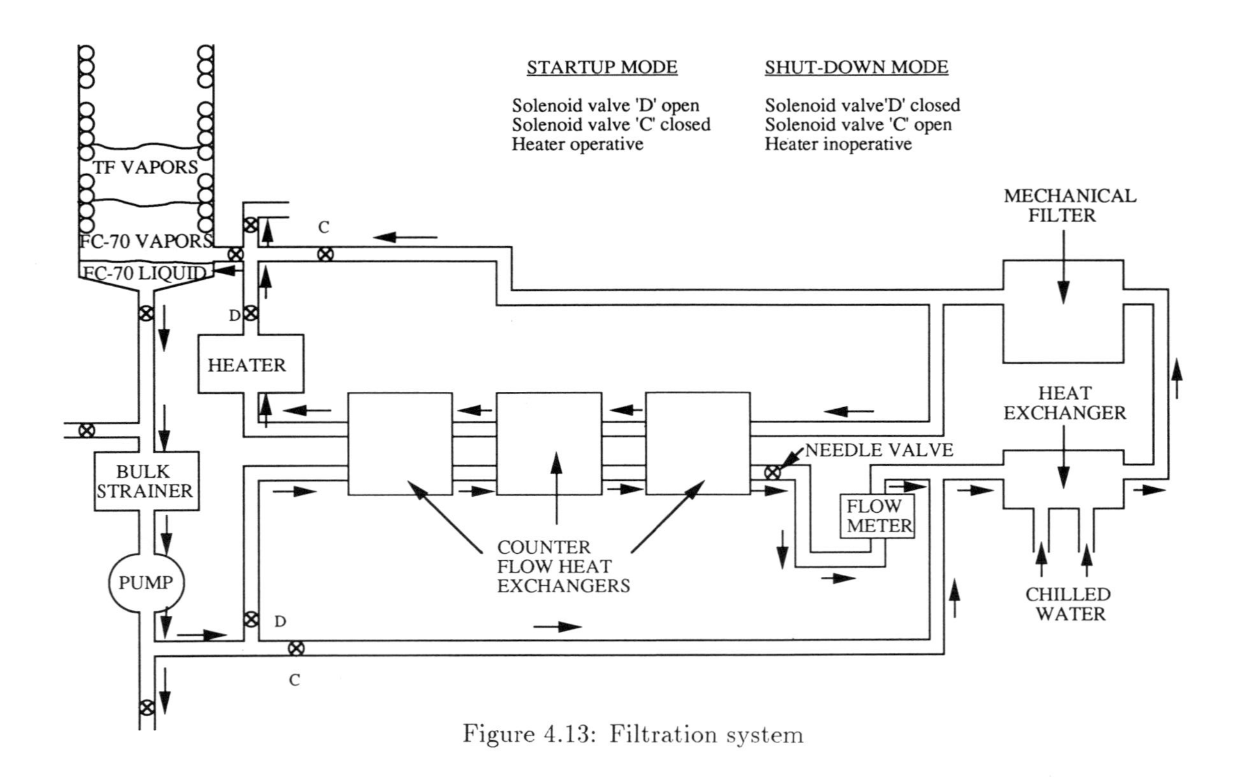

Figure 4.13: Filtration system

4.5.1 Infrared reflow soldering

Infrared (IR) soldering is a popular reflow method. First used in the late 1980s, this technique directs the energy from an infrared heat source onto the component lead site to raise the temperature of the solder joint, allowing the solder to melt (reflow), wet the component lead, and later cool down and form a reliable joint.

The main advantage of IR technology is that the temperature distribution and the maximum temperature in the system can be easily and quickly varied, giving the system the flexibility to use various solder alloys with different melting points to solder the joints continuously in the same piece of equipment. The trade-off, however, is the slow heating rate of the process.

The technique uses IR emitters or lamps to heat the joint locations in the component assembly. The solder preforms or paste on the joints absorb a part of the emitted IR energy, a radiant energy present in the invisible part of the electromagnetic spectrum. (A schematic of the electromagnetic spectrum is given in Figure 4.14.) The wavelength of these rays is greater than the wavelength of red light, and the wave band is formed next to the visible spectrum on the red side. When the molecules of the solder at the joints come into contact with these rays, they absorb the energy, increasing their natural frequency and generating heat. However, absorptivity alone does not determine the solder material temperature, which also depends on the mass, specific heat, reflectivity, thermal conductivity, diffusivity, and geometry of the joint. The generally accepted theory of IR rays assumes their generation by black-body radiation or point-source emitters. IR reflow systems must, therefore, employ a complex arrangement of various insulation materials, reflectors, emitter surface materials, and mounting arrangements to achieve effective process control.

Three popular types of IR heating systems are used in reflow systems: lamp, panel or natural-convection, and forced-convection IR systems. The selection of a particular heating system depends on the component selection, solder alloys, and other factors. These heating systems are discussed in the following sections.

Lamp IR systems

The lamp IR method of reflow soldering employs an IR lamp or quartz-encapsulated tungsten filament emitter to generate IR energy at a wavelength of about 1 to

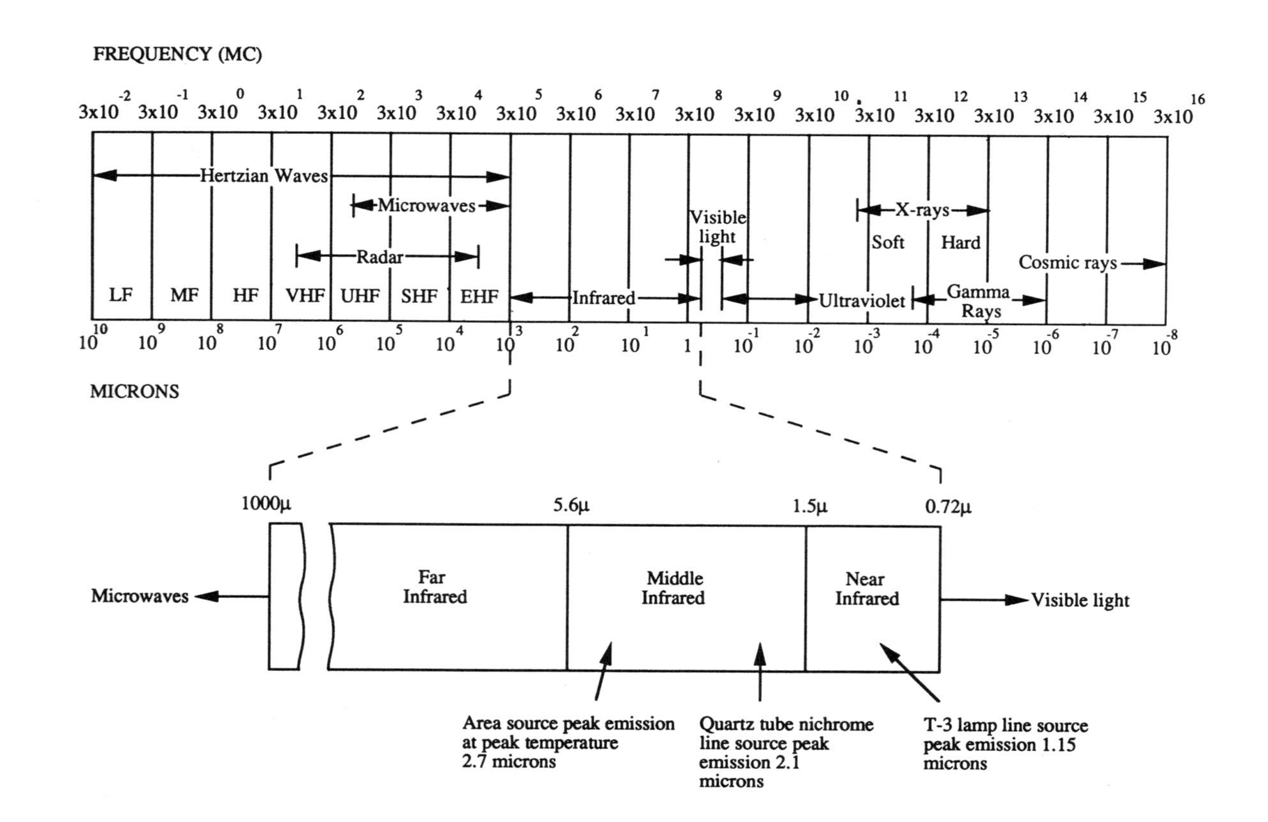

Figure 4.14: The electromagnetic spectrum

3 microns. The filaments, which consume a lot of power and operate at high temperatures, generate IR energy. The lamp IR system is able to respond rapidly to input voltage changes, quickly reaching the new equilibrium temperature. This process is very suitable for use on assemblies that require localized soldering temperatures to reflow the solder preforms.

The limitations of the lamp IR process are overheating of small components, charring of board corners due to oxidation, higher impact of color selectivity, and shadowing of closely spaced components from the incident energy. Oxidation of components, boards, or joints can be countered effectively by the use of nitrogen or a combined nitrogen and 5% hydrogen atmosphere. The use of this atmosphere increases the throughput of the process but makes it more expensive and difficult to control.

Temperature control is an important parameter in the use of this process. In lamp IR systems, convective heating is negligible because the air in the region between the joints and the emitter is transparent to the emitted spectrum of energy. The excellent flexibility of the process temperatures makes this reflow soldering method very popular among manufacturers. (Figure 4.15 is a schematic of one of the complex temperature-controlled lamp IR systems. Figures 4.16 through 4.19 show various other IR reflow machinery.) The thermal input to the printed circuit board consists of direct lamp emission, reflected emission from oven walls, re-emitted (residual) radiation, and convection heating. For effective temperature control and quick response to varying loads, the oven control system must consider each component to determine the optimum heating profile.

In a lamp IR system, close tolerance temperature control is obtained with the help of thermocouples that measure the total heat energy in the oven in real time. The readings from the thermocouple are fed back to the oven controller, which is equipped with a voltage modulator and phase-angle-fired power controls that regulate the power to the lamps. These controls allow very rapid switching, producing stable filament temperatures to compensate for loading changes. The rapid temperature drops caused by a board passing through the system (up to 40°C in a minute or less) are effectively compensated for by the controls.

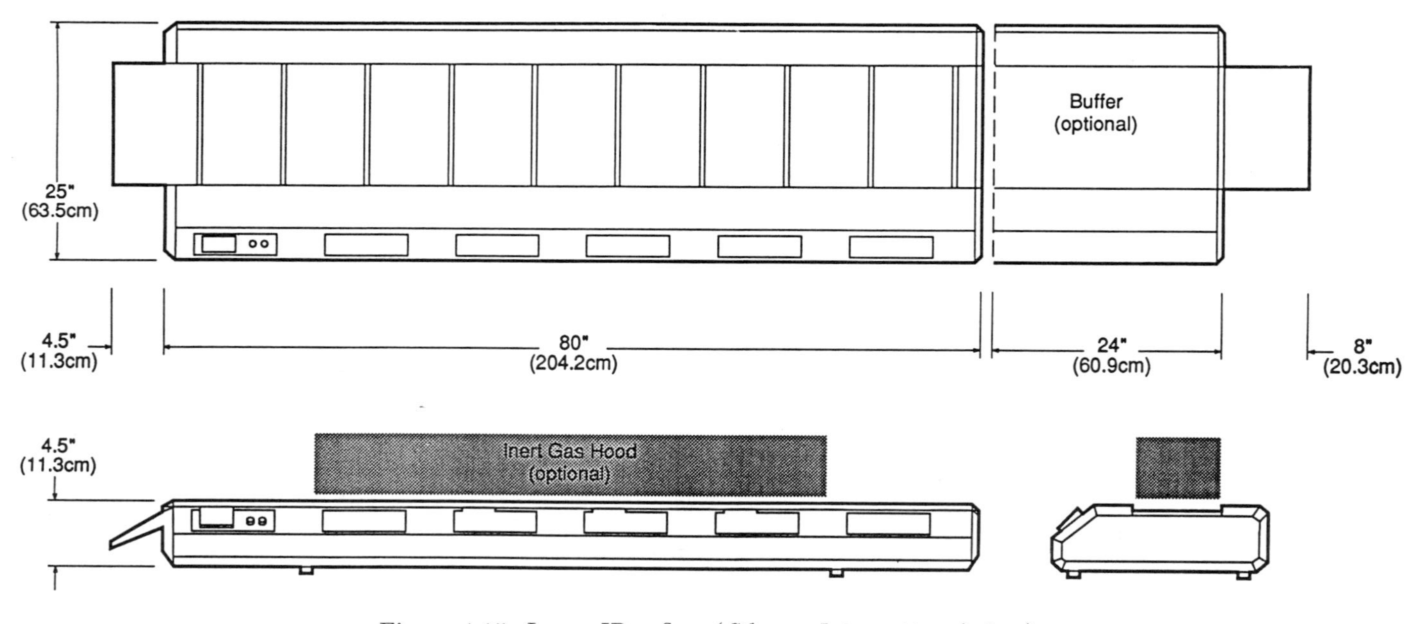

Figure 4.15: Lamp IR reflow (*Sikama International, Inc.*)

Figure 4.16: IR reflow machinery with SMT boards (*Sikama International, Inc.*)

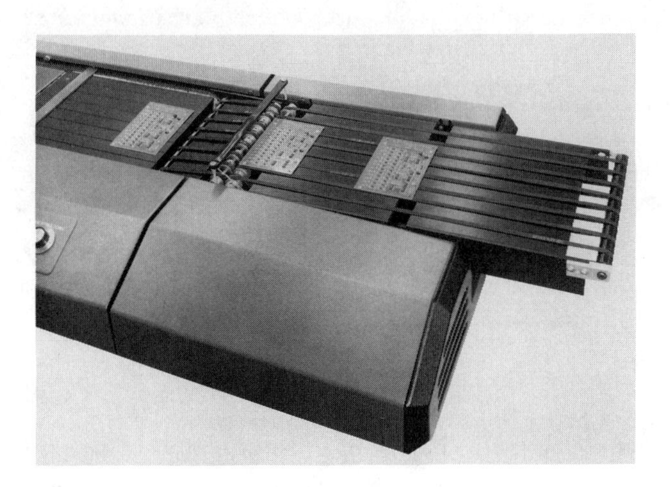

Figure 4.17: Automatic buffer on loader (*Sikama International, Inc.*)

Figure 4.18: Inert gas hood (*Sikama International, Inc.*)

Combining radiation and convection

Both convection and pure radiation heating systems have advantages and disadvantages. The advantage of a convection system is that it heats areas in a shadow region that would not get heated by a pure radiation system. Moreover, because the heat transfer is from a vapor or gas at a constant temperature, a limit is set on the maximum temperature the components can be heated to. This minimizes the chances of overheating the components on the printed circuit board.

The disadvantage with convection is that the rate of heat transfer is very low. The only ways of increasing heat transfer are to either increase the air flow or increase the temperature of the air. The former method risks moving the components; in the latter method, the advantage of reducing overheating by keeping air temperature low can be lost.

The disadvantage of radiation heat transfer is that parts can be shadowed because of tight spacing. Additionally, because there is a great temperature difference between the source and the target, uneven heating of the board and the components can occur. The advantages include precise control over the heating rate, the repeatability of the process, and heating efficiency of about 50 to 60%.

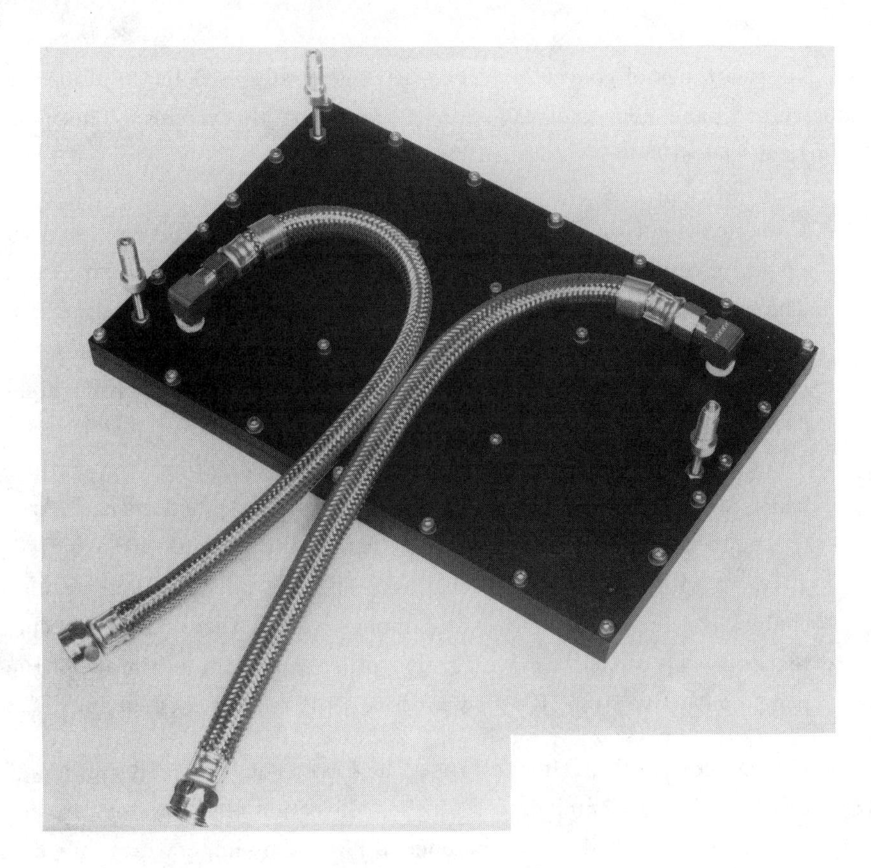

Figure 4.19: Water-cooling arrangement (*Sikama International, Inc.*)

A combination of convective and radiation heat transfer constitutes a good mix of both technologies to evenly and quickly heat the solder paste. However, to minimize the disadvantages of each type of heating, different combinations of the two are applied in different regions of the reflow chamber. The preheat zone of the reflow profile addresses the need for closely monitoring temperature increases. Most of the heat supplied is produced in this stage, and therefore radiation heat transfer is used to speed up the process. In the dryout zone, however, it is important that all parts on the board attain the same temperature before reflow occurs; therefore, convective heat transfer predominates, clearing any volatile gases produced due to radiation heating before the board experiences reflow. In the reflow zone, large amounts of heat energy are again required, and thus radiation heat transfer becomes the dominant heat transfer method once more. This also ensures that the wetting time is minimal.

The two types of convection IR systems generally used in the soldering industry are panel-convection and forced-convection IR systems. The main features of each system is described below.

- **Panel-convection IR systems.** The panel-convection IR reflow-soldering system uses waves of about 2.5 to 15 microns to generate primary heat. The secondary emission from panel surfaces provides additional heat to the component by convection from air in the enclosed chamber, thus, the transfer of heat is twofold. Convective heat transfer is approximately 60% of the total heat, while direct IR radiation accounts for the rest. The important feature of the system is the low operating temperature, which allows the conveyors to operate at much lower belt speeds. As a result, more uniform and repeatable temperature profiles are achieved, although with some loss in throughput. The controlled heating rate also enables specific component heat ramping. Thus, there are few occurrences of chip capacitor cracking. The other significant advantage of the panel-convective lamp IR reflow process is the reduction of shadows.

- **Forced-convection IR systems.** The forced-convection IR method of reflow soldering is similar to the panel-convection reflow system, except that air is circulated in the chamber to lower the temperature, thus dissipating and spreading heat by forced convection. This proves to be very useful when soldering joints of different thermal mass under high throughput conditions.

  Figure 4.21 shows a cutaway view of a forced-convection system employing IR lamps and forced convection, using exhausts and blowers for air circulation. The forced- convection IR system helps circulate air, heated to just below the temperature of the hottest region in the assembly. This air is then used to heat the cooler areas in the convection oven without overheating the areas already at a high temperature.

4.5.2 Laser-reflow soldering

The laser-reflow soldering technique uses focused energy from either a carbon-dioxide or a neodymium-doped yttrium-aluminum- garnet (Nd:YAG) laser to heat the solder joint. When properly focused with the energy of the beam

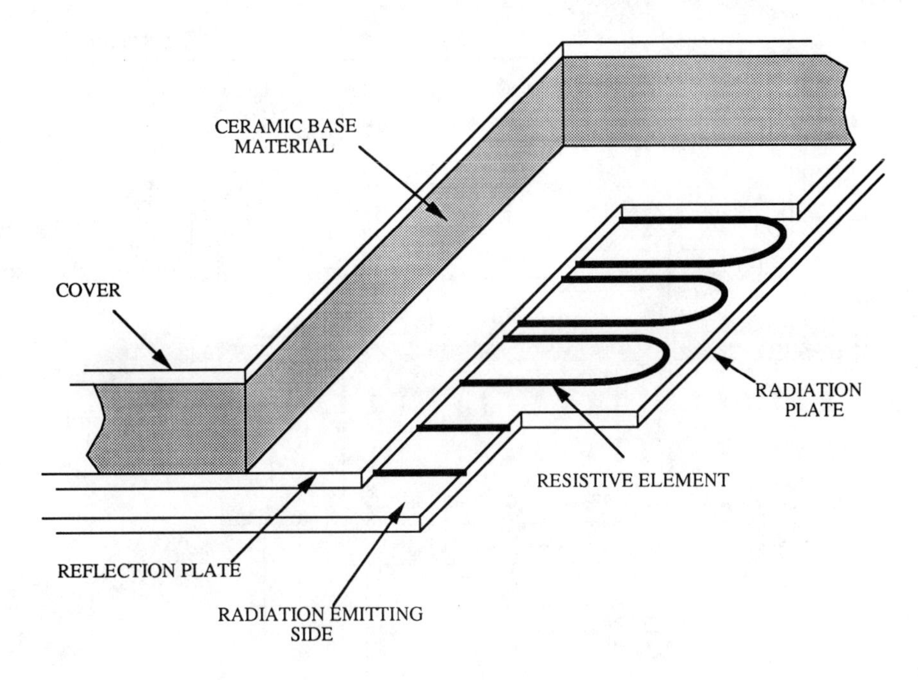

Figure 4.20: Emitters for panel-convection IR systems

efficiently controlled and the time of soldering accurately regulated, laser-reflow soldering yields very reliable joints.

The Nd:YAG laser, with a wavelength of 1.06 μm, allows more than 75% of the energy to be absorbed by the solder. In contrast, because of the CO_2 laser's higher wavelength of 10.6 μm, more than 70% of the incident energy is reflected (less than 30% absorbed) by the conductive surfaces (leads and solder) on the board. The amount of absorption largely depends on the substrate material. While the absorptive peak for the CO_2 is close to that of substrate materials such as epoxy and polyimide, allowing more energy to be absorbed by the board, the Nd:YAG laser's peak more closely corresponds to the peak of metals in device leads and the solder itself — that is, to the surfaces that actually are to be heated.

YAG lasers can focus on areas with diameters as small as 25 microns; CO_2 lasers can focus on areas greater than 100 microns. Thus, YAGs can solder very small joints. These advantages, together with other features like the simpler, less expensive optics required for production, support the use of YAGs over

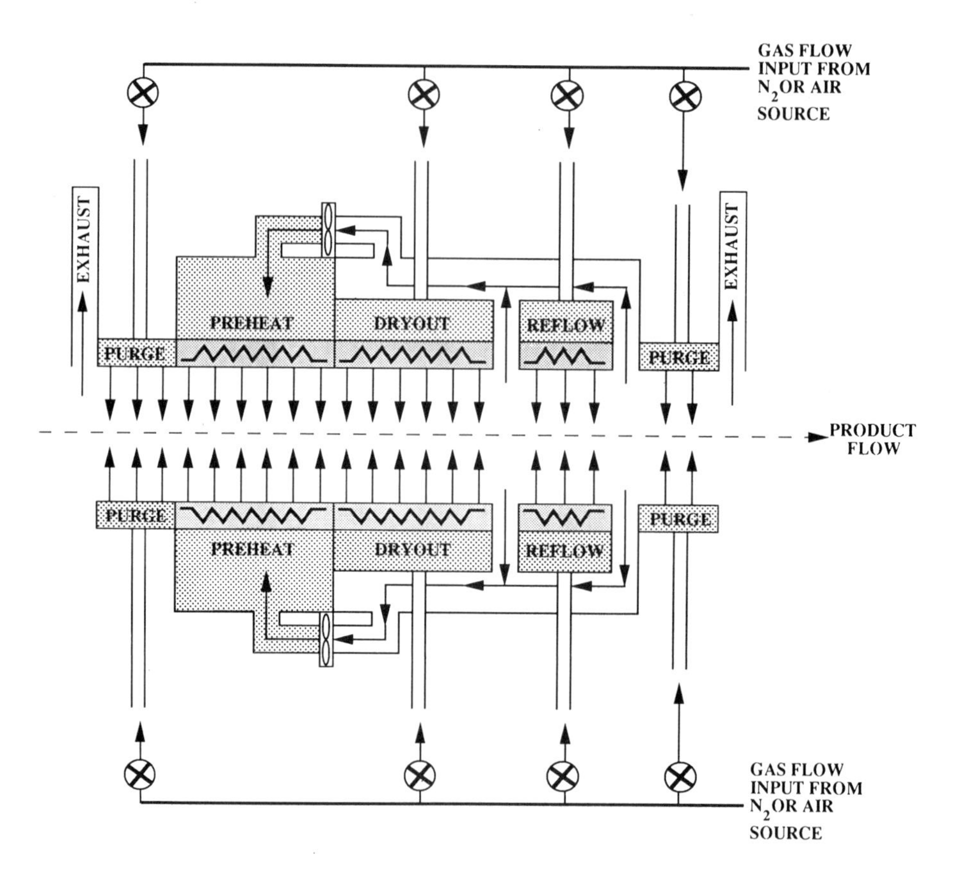

Figure 4.21: Cutaway view of forced-convection IR system

CO_2 lasers. However, applications requiring higher powers for penetration use CO_2 lasers, and YAG lasers pose potential dangers to operator eyesight.

A typical laser-soldering system consists of a laser unit with a power supply and water-cooling unit, a work base, and an enclosure for safety purposes. The YAG laser requires an arc lamp to serve as a radiation pump source; the CO_2 laser uses a specially formulated laser gas mixture. The beam path is always enclosed by the laser unit and the focusing lens. The shutter of the eyepiece microscope used to view the solder joint is triggered when the beam is activated. (Figure 4.22 shows the use of a typical laser soldering unit.)

An important requirement for laser soldering is an effective means of focusing the laser onto the solder joint. The YAG system employs coaxial viewing

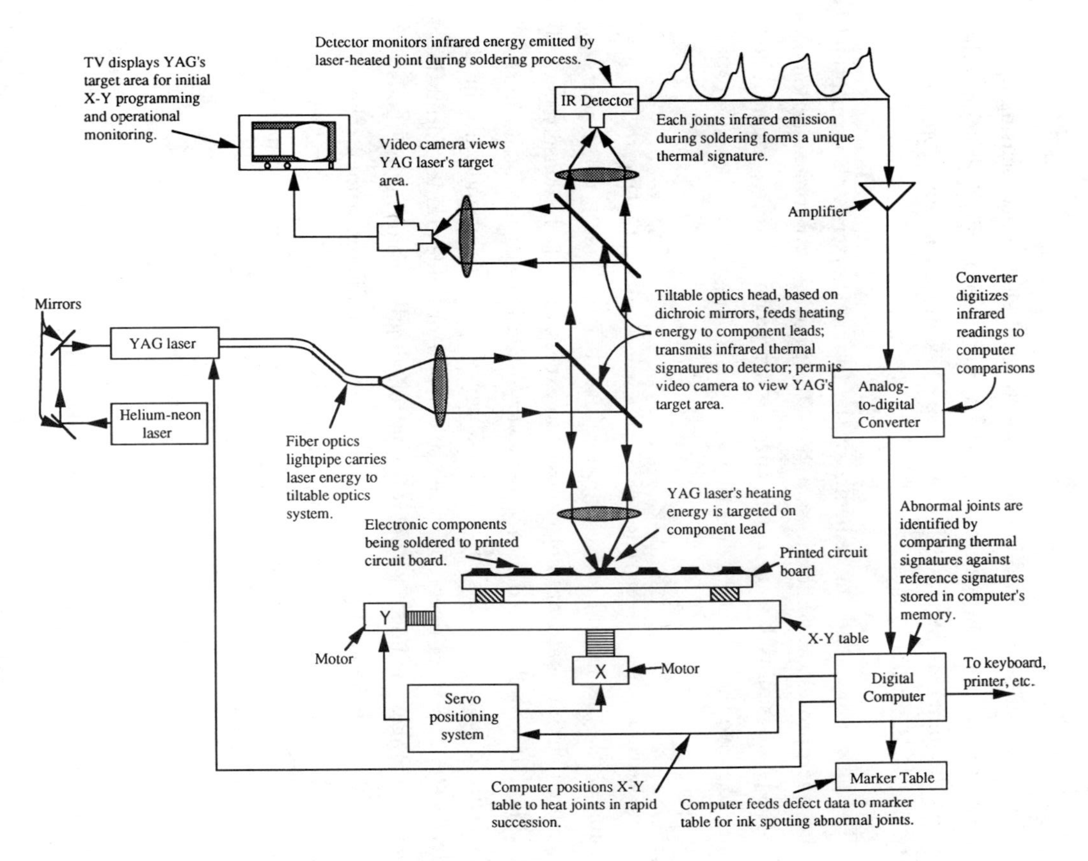

Figure 4.22: Optics for laser-reflow soldering

and focusing for this purpose. The CO_2 beam, on the other hand, uses an off-axis binocular or closed-circuit television viewing system or a helium-neon laser-spotting beam.

Focusing can be achieved in one of two ways. Usually, the focusing optics for the beam (i.e., the objective lens and the mirrors used to direct and focus the beam on the joint) are fixed, while the workpiece is moved on a microprocessor-controlled X-Y motion table.

The laser power and the sequence of the timing shutter is also controlled by the microprocessor. The alternative is to keep the workpiece stationary while the beam is moved, using small, current- driven microprocessor-controlled motors to move the reflectors directing the laser beam. This process is called galvo-optics.

The advantages of laser-reflow soldering are significant. Due to the fine focusing of the beam and short process time (0.3 to 1 sec.), other parts of the printed circuit board are not exposed to excessive heat. In densely packed assemblies, laser-reflow soldering proves very effective because of its narrow and accurate focusing abilities. In the absence of excessive moisture in the solder paste, the laser does not cause solder-ball or solder-bridge formation. Other advantages include high process control, simpler mounting methods for components, and the ability to mount components at different times. The limitations of laser-reflow soldering are low throughput rates, high capital costs for equipment procurement and set-up, exacting safety requirements, and complicated process controls.

4.6 Trends in reflow soldering

Reflow soldering has become one of the predominant methods of soldering used in industry, primarily because of its usefulness and convenience over other methods for fine-pitch technology, which involves soldering dimensions of less than 50 mils. Apart from fine-pitch reflow, some newer technologies involving the use of reflow soldering have come into use. The basic concern in reflow soldering is to apply the right amount of heat at the right time at the right place. Most of the technologies discussed here address high-quality soldering.

4.6.1 Microflame-reflow soldering

Microflame-reflow soldering combines the soldering iron and laser-soldering approaches. The concept is that good thermal contact between the soldering iron and the solder point is not essential for efficient heat transfer.

The microflame soldering head is designed to be precisely controlled to a hundredth of a second. A constant reproducible energy supply is provided by a multi-cell gas generator. The head can be robotic.

The machine is generally supplied with 100-liter/hr. or 200-liter/hr. gas output capacity. The multi-cell gas generator is the source of residue-free, clean gas. The generator has a high operating reliability through water electrolysis, and allows up to three times the normal output of a single-cell generator with 85% less heat loss. In Figure 4.23, the gas-generator system is shown set up in the fully programmable micro-flame soldering and soldering iron robots. Figure 4.24 shows a modular semi-automatic soldering work station for applications not requiring the flexibility of a robot. The equipment can be used for both microflame and precision-microsoldering technology with a soldering tip.

The advantages of the process include:

- Processing time is half that of the soldering-iron technique.
- Soldering quality is extremely consistent.
- Cleaning between soldering cycles is obviated.
- Less maintenance and servicing are required.
- The price/performance ratio is much more favorable than that of other techniques.
- Process reliability is greater than in laser soldering, because the microflame applies a constant amount of heat to the solder point, preventing dry (cold) joints.
- The flame is directed away from the board when it moves between two soldering positions to ensure that the area around the solder joint does not heat up.

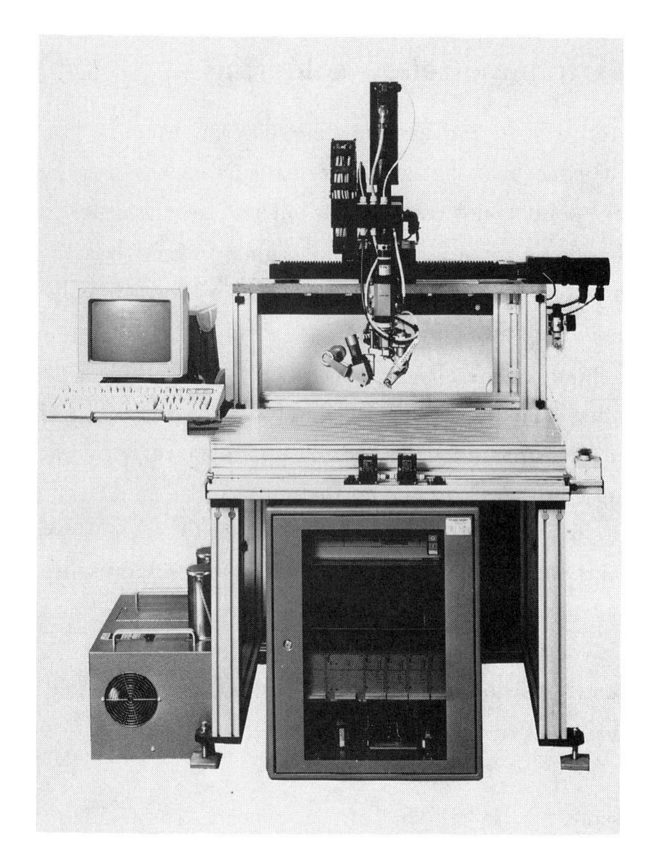

Figure 4.23: Microflame reflow-soldering system

4.6.2 Optical-fiber reflow soldering

YAG lasers to localize the heat supplied to the solder paste and preforms have been limited to a few high-tech applications because of cost considerations [Kobayashi 1991]. An alternative local reflow technique uses optical-fiber technology. Optical fibers direct heat from a radiant source, such as a lamp, to specific solder sites so that components and the surrounding board area are not affected.

One main advantage of optical fiber reflow is that the heat source never touches the components being soldered, as in manual soldering with a soldering iron. Therefore, there is no danger of components being displaced. Moreover, in the soldering-iron technique, heating depends on the amount of contact be-

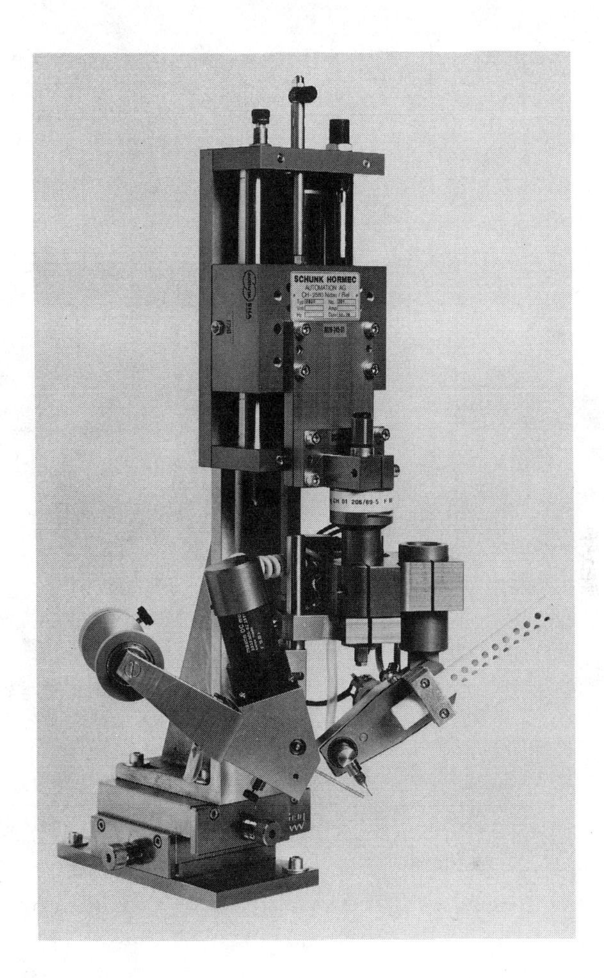

Figure 4.24: Modular microflame reflow-soldering station

tween the soldering-iron tip and the solder paste or preform. The technique is manual, and therefore variation is considerable and non-uniform soldering results. Optical-fiber reflow soldering eliminates variations in soldering quality because the energy is transferred by radiation, not by conduction. Another advantage is that the solderer does not have to worry about the temperature increase of the workpiece, because the radiation heat transfer to it from the heat source can be ignored. Finally, the response time of the lamp (generally xenon) is much shorter, so process control is excellent.

Optical-fiber reflow-soldering equipment consists of a xenon lamp with a short arc; light converges into the receiving end of the optical-fiber bundle with

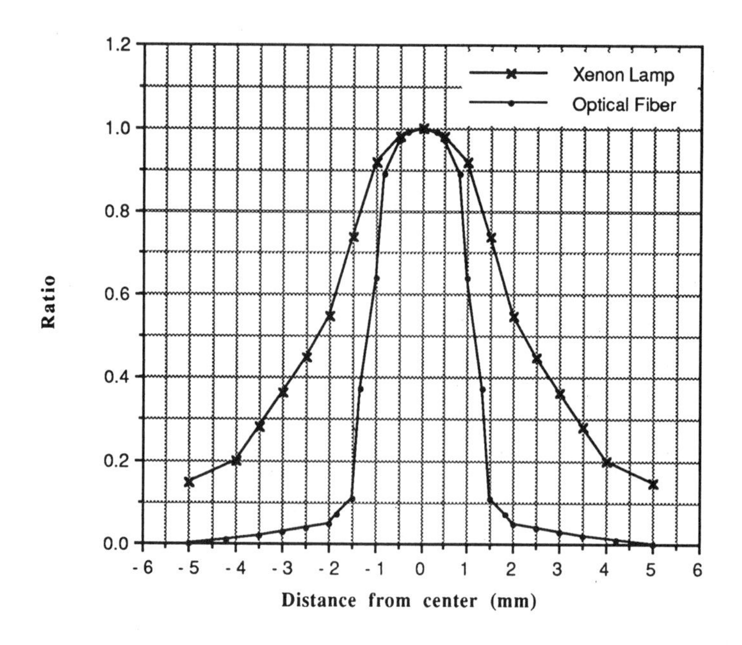

Figure 4.25: Energy distribution curve of the xenon lamp

the help of an ellipsoidal mirror. The optical-fiber bundle then carries the near-infrared-to-ultraviolet light to the solder site by a series of internal reflections within the fibers. The bundle of fibers is very flexible and can be moved to different parts of the board either manually or with a robotic arm. At the other end of the bundle, the light again passes through an irradiant lens and focuses on the solder site.

Optical fibers restrict the high-energy portion of the xenon energy curve very close to the center of the fiber (the high-energy band is about the diameter of the optical fiber, as shown in Figure 4.25). Additionally, the diameter of the light spot on the printed circuit board can be varied without any change in the intensity of the incident radiation by adjusting the current supply to the xenon lamp and, thus, changing the amount of energy radiated by it.

Probably the greatest advantage offered by optical-fiber technology is the possibility of its use with a robotic arm controlling and coordinating the precise movements required to position the heat source on the board. The required position may even be on the underside of the board, allowing components on both sides to be soldered simultaneously. The robot need not be very complex; an ordinary cartesian or point-to-point motion robotic arm is all that is required.

4.6.3 Chip-level reflow-soldering applications

Fine-pitch manufacturing calls for a tighter control on tolerances and an investment in newly emerging soldering technologies, along with the incorporation of sophisticated equipment. For high volume assembly, placement machines need electronic vision to achieve the desired placement accuracy. The general trend is to use hot-bar reflow-soldering technology or fine-tip soldering irons to individually attach the fine-pitch packages. Some of the improvements in existing systems required for fine-pitch applications of reflow soldering are outlined in the following sections.

Fine-pitch dispensing

Existing state-of-the-art technologies can dispense solder in different-sized dots, using positive displacement pumps with programmable shot size [Cavallaro 1991]. Solder dots can be dispensed at 50- or 25-mil-pitch footprints with these dispensers. To achieve such accuracy, rotary positive-displacement pumps are used with constant pressures of 4 to 10 psi to keep the paste flowing. An Archimedes-style auger screw dispenses the solder. Control of the amount of liquid dispensed is precise enough to measure 5-millisecond increments at speeds of over 16,000 dots/hr. when dot sizes range from 12 to 50 mils. Needles specifically designed for accuracy do the dispensing.

The main problem in dispensing systems is developing a sensor system capable of recognizing the amount of solder to be applied at a particular point, depending on the type of connection to be established. In general, the dispense height must be maintained at half the diameter of the needle to ensure proper dot formation. Z-axis compensators are used to avoid "tailing" of the dots, a tendency of viscous materials to leave excess material on the board. Various types of Z-axis compensators are listed below.

- **Mechanical Z sensors.** Mechanical Z sensors use a precision slide to help position a needle above the board. The needle never touches the board. Instead, a mechanical foot moves down on the board, presses on it to dispense the solder paste, and moves up again. The mechanical Z sensor works at a rate of 16,000 dots/hr. making it the fastest available.

- **Programmable Z compensators.** Programmable Z compensators can maintain several different heights of dispensed solder. A touch probe is used to find the dispensing site and set the dispensing height. Therefore, large as well as small integrated circuits can be securely bonded to the board, using the same dispensing equipment. The speed of the process is about 8,000 dots/hr.

- **Non-contacting Z compensators.** Non-contacting Z compensators are used when the board surface is too delicate to withstand even a few grams of pressure. Laser vision is used to measure the Z-axis height before dispensing, to make adjustments, and to reset dispensing height. A considerable amount of time is required to look for the dispense position with a laser-sensing device. The non-contacting Z-compensator method works at typical speeds of 7,500 dots/hr.

The vision system employed is an important consideration in automated dispensing equipment when accurate positioning of the dispensing needle is required. The most sophisticated are gray- scale vision systems using color monitors, light-emitting diode illumination, and closed-circuit device cameras to achieve pattern recognition from any component on the board.

To increase productivity, automatic dispensing systems are integrated with pick-and-place robots. Speeds of up to 25,000 dots/hr. can be achieved with such systems.

Preheaters

The preheaters for fine-pitch manufacturing must allow better flow of epoxies and uniform setting of the epoxies, and avoid air-pockets. In one process, the boards are preheated during encapsulation to 100°C and then moved to a dispense station, where a lower-temperature IR heater is used to maintain temperature while the epoxy is applied. X-ray photography shows that the number of defects and subsequent rejects is reduced with the use of such processes.

4.7 References

Cavallaro, Kenneth J. Alternate Techniques for Fine-Pitch Dispensing. *Circuits Assembly* (May 1991).

Engel, Jack. Solder Paste Dispensing on Micro-Hybrid Substrates for Surface Mount Applications. *Hybrid Circuit Technology* (July 1991), 31-34.

Enterkin, Robert. Equipment/Material Synergism Key to Solder Paste Printing. *Electronic Packaging & Production* (May 1991), 136-138.

Hodson, Thomas L. Spray Fluxing for Today's Soldering Processes. *Electronic Packaging & Production* (January 1992), 47.

Keeler, Robert. Starting Out with Solder Pastes. *Electronic Packaging & Production* (April 1989), 56-59.

Kobayashi, Makoto. Local Soldering with Light Beam Using Optical Fiber. *Proceedings of the Technical Program* NEPCON West (1991).

Manko, Howard H. *Solders and Soldering.* New York: McGraw-Hill Publishing Co. (1992).

Paavola, Bob. Revisiting Vapor Phase. *Circuits Assembly* (October 1990).

Research, Inc. Technical Literature (1992).

Rowland, Robert J. Screen Printing Surface Mount Adhesives. *Proceedings of the Technical Program* NEPCON West (1990).

Zarrow, Phil. Controlling IR Reflow in SMT Soldering. *Electronics Manufacturing Engineering,* 7(1), (First Quarter 1992).

Chapter 5

Cleaning and Contamination

David A. Curtis
Curtis Associates, Inc.
Concord, MA

5.1 Situational analysis

5.1.1 The challenge

The soldering process tends to create byproducts that contaminate the printed circuit board. Many of these contaminants create damaging corrosion. In general, a contaminant is any material — not just those that cause corrosion — that can degrade the printed circuit board's performance to unacceptable levels. Contaminants may be on the printed circuit board's surface or embedded in it, chemically bonded to the surface, absorbed in a porous surface, or inadvertently encapsulated into the board structure. This chapter focusses on cleaning contaminants created by the soldering process, although printed circuit board fabrication and solder-masking processes can also create contaminants.

The principal source of soldering contaminants is the flux used during soldering. However, other sources include body oils, talcs, perfumes, dried skin, and dandruff from workers handling the boards; oils used by component placement machines; adhesives; solder balls; and the atmosphere. To reduce the chance of corrosion and increase the reliability of the finished product, the printed circuit board must be cleaned after soldering; if it is soldered several times, repeated cleaning may be necessary.

As discussed in section 5.3.2, fluxes are used during soldering to clean the surfaces to be joined, particularly to remove oxides. They also must keep the surfaces clean during the solder process. The flux residue is the main source of contamination. At first sight it appears an aggressive flux is best, because it efficiently cleans the surfaces to be soldered. Unfortunately, aggressive fluxes also tend to leave aggressive contaminants that reduce the reliability of the printed circuit boards. These fluxes require careful cleaning. Good flux selection balances effectiveness and ease in cleaning the surfaces to be soldered.

A second important property of fluxes is their tack — the ability to hold a component in place during the solder process. Unfortunately, tacky fluxes also attract particulate contamination. Desirable characteristics of a flux, all of which should be available at a reasonable cost, include:

- environmental safety;
- low toxicity and flammability;
- sufficient activity to give good soldering performance;
- minimal corrosion from both flux and residue;
- excellent cleanability after soldering;
- good appearance with very little or no visible residue;
- a boiling point above room temperature;
- stability to provide good shelf life and reproducibility during use;
- good rheological properties to promote ease of application;
- good flux tack to hold components in place but no residue tack;
- compatibility with components and equipment.

Solderability is usually monitored using wetting balance tests (see Chapter 6); appearance is usually monitored visually. Contamination and corrosion performance are monitored by their impact on printed circuit board reliability. printed circuit boards should be subjected to accelerated-life tests to determine their reliability.

As a result of these sometimes conflicting requirements, many types of fluxes are available, each type placing different demands on the cleaning process. The most commonly used fluxes are listed below [Prasad 1989].

- **Organic acids (OA):** Organic acids with good activity leave contaminants containing polar ions, which are quite corrosive; these contaminants can be cleaned using water, an excellent polar solvent. If the flux formulation has a low solids content (that is, a low concentration of active ingredients), the aggressiveness of the flux is tempered, but cleanability is improved. OA fluxes have not been commonly used in solder pastes needed for surface-mount assembly. Their main application is in wave soldering.

- **Superactivated fluxes:** Superactivated fluxes — super-rosin-activated (SRA) and synthetic-activated (SA) — act like OA fluxes but are less corrosive. However, they require a non-aqueous solvent for cleaning. Like OA fluxes, SRA fluxes have not been commonly used in solder pastes.

- **Rosin fluxes:** Rosin fluxes fall into four main types: rosin (R), rosin mildly activated (RMA), rosin-activated (RA), and RA for military uses. RMA fluxes are frequently used in solder pastes for reflow soldering. Both RMA and RA fluxes are used to solder SMT components. The activation referred to above is the addition of a halide to increase the otherwise weak activity of the flux. The greater the activation, the more important it is to clean the printed circuit board after soldering. In some applications, R and RMA fluxes are not cleaned but their tackiness, valuable in holding components in position before soldering, can create contamination problems because dirt adheres to the flux.

This traditional classification of fluxes, widely understood by many manufacturing engineers, is also consistent with the nomenclature used by many flux manufacturers. However, several years ago, the Institute for Interconnecting and Packaging Electronic Circuits (IPC) issued standard IPC-SF-818, which classifies fluxes by their level of activity (low, medium, or high), not by their composition.

Polar		Non-polar	
1.	Acid fluxes	1.	Gums
2.	Etchants used in circuit definition	2.	Hand lotions
		3.	Hair sprays
3.	Fingerprints	4.	Makeup
4.	Perspiration	5.	Oils and greases
5.	Plating salts used in board fabrication	6.	Rosin fluxes

Table 5.1: Typical contaminants found on printed circuit boards

5.1.2 The impact of changing technology

The basic principles of cleaning electronics have been known for many years [Agnew 1974]. Contaminants are either ionic (or polar) or non-ionic (non-polar) (see Table 5.1). Polar contaminants cause corrosion if they are left on a printed circuit board. However, tacky non-polar contaminants trap both polar and non-polar dirt; if they catch the former, then corrosion occurs. Non-polar contaminants can also create dielectric layers on bonding pads and test points, thereby preventing circuit operation or inhibiting testing with probes.

Non-polar contamination should be removed first, because polar cleaning fluids cannot penetrate non-polar dirt. The most commonly used solvents for non-polar materials are organic fluids. The selection of the best organic solvent depends sensitively on the kinds of plastics used in the electronics assembly. After the non-polar contamination has been cleaned off, a polar agent like water, alcohol, or acid should be used to clean off polar contamination.

Rather than using a two-step process, many companies prefer to use a blend or emulsion of polar and non-polar solvents in the form of an azeotropic mixture that maintains the ratio between the components even as it evaporates. Unfortunately, many of these azeotropes contain chlorine compounds. The most important change in cleaning technology is the recognition that chlorinated solvents impact the environment. This topic is discussed further in the next section.

Of course, some contaminants, most notably particulates, such as solder balls, chips of printed circuit board laminate, and some forms of dust, are not soluble in the commonly used polar or non-polar solvents. Then, cleaning relies on the motion of the solvent around the printed circuit board to first free contaminants and then sweep them away. Other important changes in technology include:

- the use of finer, more closely spaced, leads on component packages;
- an increase in the size of integrated circuit packages;
- the lower stand-off height under packages; and
- the greater use of adhesives to attach components to printed circuit boards.

The first three changes all lead to the same problem — contaminants trapped under components. Thus, techniques are needed to facilitate entry of the cleaning solvent under the components and to remove the solvent with the contaminants after the cleaning is finished. The fourth point, the greater use of adhesives, just creates another source of contaminants with different chemical characteristics and, if the adhesive curing process is not properly controlled, another place where contaminants can be trapped. A 28-pin dual-in-line package (DIP) of the type used in conventional through-hole printed circuit boards has an area of about 0.75 sq. in. (452 sq. mm) and a stand-off height of 0.036 in. (0.9 mm). In contrast, an 84-lead plastic-leaded chip carrier (PLCC) has an area of 1.4 sq. in. (900 sq. mm), a stand-off height of 0.025 in. (0.6 mm), and small-outline integrated circuits (SOICs); smaller areas have even lower stand-off heights (see Table 5.2).

printed circuit board designers should consider cleaning when selecting components and laying them out on the printed circuit board. Some basic guidelines are listed below.

- Plated through-holes that exit under a component should be avoided. They can vent flux under the component that is very difficult to remove later. If the layout demands one or more holes, then a solder mask should be used on at least one side of the board to close off the hole.
- Solder masks should be tested to ensure they adhere well throughout the mass-reflow solder process.
- The components should be laid out so leads and solder joints present the smallest obstruction of flux and solder, with both covering the board and

Component	Area (sq. in.)	Standoff height (mils)
PLCCs:		
20 leads	0.15	25
44 leads	0.45	25
68 leads	0.90	25
84 leads	1.40	25
SOICs:		
14 leads	0.10	8
28 leads	0.25	11
LCCs:		
20 leads	0.25	3
44 leads	0.45	3
68 leads	0.90	3
84 leads	1.40	3

Table 5.2: Typical areas and stand-off heights for SMT components

then draining off it. If, as in most wave-solder systems, the board is at an angle to the horizontal, it should be oriented like DIPs and SOICs, with their leads parallel to the direction of travel of the board. In contrast, passive components should be placed perpendicular to the direction of travel.

5.1.3 The Montreal Protocol

Unfortunately, many common solvents, such as chlorofluorocarbons (CFCs) and Freon 113 (trichlorotrifluoroethane), are often used in cleaning electronics; perchloroethylene and methyl chloride are now known to damage the environment. The property that makes them attractive as cleaning agents is also their downfall in the environment; they are very stable when they reach the ozone layer in the stratosphere, and this stability allows them to destroy large amounts of ozone before they dissipate. Ozone destruction reduces the earth's natural protection against ultraviolet radiation and contributes to the greenhouse effect.

In 1987, twenty-four countries attended a conference organized by the United Nations Environmental Program in Montreal, Canada. They signed a protocol agreeing to first reduce and then eliminate the use of CFCs. This protocol has since been amended at a second conference in London and its terms implemented through many regulations issued by national, state, and local agencies. Although the electronics industry is a relatively small user of CFCs, responsible for about 12% of the total use in 1987, good corporate citizenship demands the elimination of CFCs from printed circuit board cleaning processes. In any case, CFCs will not be manufactured after 2000 and hydrogenated CFCs (HCFCs) will not be manufactured after 2020.

These dates have changed several times since the Montreal meeting, and some countries have adopted much more stringent regulations. For example, the dates adopted by the German government are 1995 and 2000, respectively. On February 11, 1992, President Bush instructed the EPA to limit the production of CFCs, halons, methyl chloride, and carbon tetrachloride after December 31, 1995, to the levels needed for essential medical uses and for servicing existing equipment. He also indicated that the EPA will consider accelerating the phase-out of production of HCFCs to 2005 and might bring methyl bromide into the coverage of the 1990 revisions to the Clean Air Act.

5.1.4 Major alternative cleaning techniques

The major alternative cleaning materials are detailed below.

- **HCFCs:** HCFCs contain less chlorine than the other systems discussed here. Their use is also limited by the Montreal Protocol, but many view them as a good interim solution to the problem. Their main advantage is that the equipment changes required are relatively minor.

- **Semi-aqueous systems:** Semi-aqueous systems use CFCs or other solvents and water, usually in a self-contained system.

- **Aqueous systems:** Aqueous systems use water, water with a detergent, or water with a saponifying agent that decomposes many of the residues, making them easier to wash away.

- Terpenes, naturally occurring chemicals with good cleaning properties, are often used in semi-aqueous systems. Unfortunately, they have a strong odor and are combustible.

- New fluxes and solders do not need to be cleaned.

Each alternative requires changes in the cleaning equipment used, and many require different fluxes and solders. Some of these changes can be threatening to some companies.

Fortunately, there are enough success stories to confirm that manufacturers can change from CFCs to other cleaning processes and achieve good process yields and economics. Test programs sponsored by the EPA, DOD, IPC, and the Industry Cooperative for Ozone Layer Protection (ICOLP) have demonstrated that several new cleaning agents perform to acceptable standards. Manufacturing engineers can assure their managements that there are good ways to replace CFC cleaners in production. The next sections describe some of these CFC-free cleaning techniques in more detail. The focus is on commercial production of SMT and mixed SMT-through-hole printed circuit boards. (For a discussion of the principles of cleaning using CFCs and other materials on the embargoed list, consult Chapter 11 of Capillo [1990]).

5.2 Desirable solvent properties

Besides the major challenge of finding solvents that do not damage the environment, there is the additional challenge of finding solvents that do not damage the components and printed circuit boards. These requirements must be met without degrading cleaning efficiency. The major physical parameters used to characterize solvents include those listed below.

- **Surface tension.** This parameter controls a solvent's ability to spread over the printed circuit board, its components, and the contaminants on the board — the wetting process. Without wetting, it is unlikely the solvent will be able to break the chemical bonds between the contaminant and the printed circuit board and components. Typically, the lower the surface tension of the solvent, the better the wetting. However, the surface-free energy of the printed circuit board is also important; this parameter can vary by a factor of three, from 19 dynes per centimeter for polytetrafluoroethylene polymer to 61 dynes per centimeter for urea-formaldehyde resin. The value for epoxy glass is 38 dynes per centimeter. If the surface tension of the solvent is greater than the surface-free energy of the printed circuit board, it will not wet. Because the surface tension of water is 78 dynes per centimeter, it does not wet common printed circuit board materials. Whenever water is used as a cleaning agent, surfactants that improve its wetting capabilities must be added.

- **Capillary action.** This is the ability of a solvent to penetrate into narrow spaces, especially under components. At first sight, a solvent with a high surface tension is preferred, because it has a higher capillary action and can penetrate more readily than solvents with lower surface tensions. However, once in the narrow space, such solvents may be difficult to remove; once again, a lower surface tension appears better than a higher value. Also, viscosity is an important controlling parameter for capillary action — the lower, the better. This trade-off among surface tension, viscosity, and capillary action is probably not as important as it appears, because most cleaning systems use mechanical means to encourage the solvents to penetrate narrow spaces.

- **Solvent density.** High-density solvents are preferable to low-density solvents for several reasons. First, they reduce losses into the atmosphere above the cleaning bath; second, the solvent flows over boards under the influence of gravity; and third, when used in a spray cleaner, the droplets have higher kinetic energy to dislodge contaminants physically bonded to the printed circuit board surface.

- **Boiling point and flash point.** The higher the boiling point or, alternatively, the lower the vapor pressure, the lower the fugitive emissions from the cleaning systems. Thus, the use of heated solvents is preferable, because they tend to have better cleaning efficiencies: as the temperature increases, the solubility of contaminants in the solvent increases and the surface tension decreases, encouraging wetting. Another advantage was revealed in the overall printed circuit board assembly process: because it is usually best to clean warm printed circuit boards before the flux hardens, it is helpful to use warm solvents — cold solvents promote flux hardening. Obviously, at no time should the solvent be heated to temperatures close to the flash point. Terpenes are an example of useful solvents whose temperature must be carefully monitored during the cleaning process because of their relatively low flash point.

- **Solubility.** The higher the solubility of the contaminants in the solvent, the more efficient the cleaning process will be.

- **Other properties.** Solvents should meet all of the user safety and environmental requirements set out for fluxes at the beginning of the chapter.

5.3 CFC-free cleaning techniques

Three established, but not necessarily unique, cleaning techniques that have received support from one or more major users are given below.

5.3.1 Semi-aqueous cleaning

The basic principle of semi-aqueous cleaning is similar to that expressed in Section 5.1.2; namely, two solvents are used in succession to clean the printed circuit board, with each solvent optimized for best performance of part of the process. The first solvent can be either a single compound designed to remove the non-polar contaminants or an azeotrope designed to remove both polar and non-polar contaminants. The second solvent is water, used to remove polar contaminants and to remove the residue left by the first solvent. The final step in the process is to dry the printed circuit board to prevent corrosion caused by any remaining water. Figure 5.1 schematically represents a typical in-line semi-aqueous system. Batch systems are also available, but usually have a lower printed circuit board throughput rate.

Semi-aqueous cleaning requires specialized cleaning equipment to contain the solvents and their vapors, to prevent the solvents from contaminating each other, to recycle solvents where possible, and to ensure that the system works efficiently [Hayes 1990]. Equipment manufacturers may need to use additional energy to improve the cleaning power of the system. Just dipping the printed

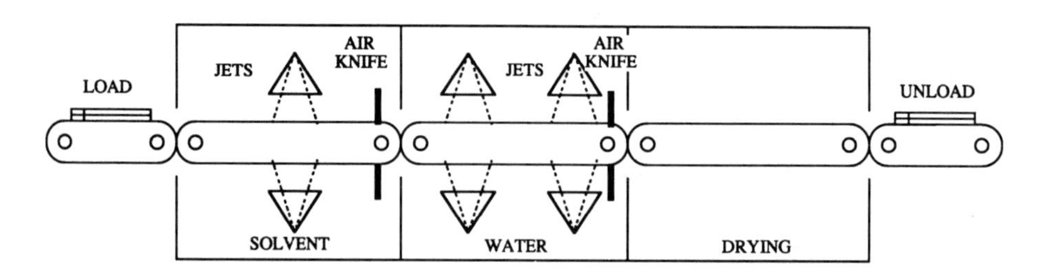

Figure 5.1: The principles of a semi-aqueous cleaning system

circuit boards into the solvents may not be enough to wet them completely, especially under surface-mount components. In any case, even if the solvent does penetrate under the components, the viscosity of typical solvents increases as they dissolve contaminants, making them difficult to remove. This problem is exacerbated if non-soluble particles are entrained in the solvent and wedge under components. Many semi-aqueous cleaning systems heat the solvents to improve their cleaning power.Heating can be very effective, as long as the solvent vapor pressure is low and its flash point is high. Heating also tends to lower a fluid's viscosity, increasing the likelihood of penetration under components. However, heating alone may not be enough to ensure adequate cleaning.

The other major way to add energy to the solvent is to increase its kinetic energy using jets, spinners, or ultrasound. Several different jet designs have been used, as the next section describes. Usually pressures in the range of 20 psi to 50 psi are sufficient. Spinning systems can hold the printed circuit board under a jet or immerse it in the cleaning fluid as do ultrasound systems. The latest systems rapidly change the frequency of the ultrasonic waves so they form a beam that scans across a printed circuit board, ensuring that the cleaning fluid is forced under all the components on the assembly. For many years engineers have expressed concern that ultrasound could damage the wire bonds in integrated circuits, but the results of recent tests show it does not [Murray 1990].

While the solvents most often used are non-chlorinated, non-halogenated hydrocarbon, terpenes are also a commonly used class. Terpenes are certified by the FDA as Generally Regarded As Safe (GRAS). These hydrocarbons are also considered biodegradable. However, some have an intense odor and most are fire hazards. The vapors and fine droplets created during spraying are flammable. The solvents should be treated as National Fire Protection Agency (NFPA) class A hazards. Fortunately, the liquids have low vapor pressures, with flash points ranging from 47° to 82°C. For these reasons, semi-aqueous cleaning systems must be self-contained, with flame detectors (to prevent fire entering the system from outside), carbon dioxide fire suppression systems, and temperature controllers to maintain the solvent at a temperature well below the flash point but high enough to provide adequate cleaning. These solvents are also regarded as volatile organic compounds (VOCs) and, as such, are subject to EPA and local regulations if discharged into the atmosphere. However, in some jurisdictions, the vapor can be discharged without violating the regulations, because such small quantities are created; this is a direct result of the low vapor pressure and careful temperature control needed. Companies considering using these cleaning systems should check their local codes carefully.

As stated above, the second cleaning step uses either de-ionized water or, in some areas with soft water, municipal water. The printed circuit boards are then passed under air jets to remove the remaining water and heated to drive off any trapped water or water vapor. These steps are easily performed, but some companies may find the cost of energy for drying a burden.

Water must be kept out of the first solvent to maintain its cleaning power. However, some of the other solvent inevitably ends up in the water, so that it cannot be discharged directly into the sewers. Many systems have self-contained water-recycling systems that separate waste water into two parts: used solvent containing flux and other residues and water (see Figure 5.2). The contaminated solvent is either burnt as a fuel or returned to the original manufacturer for recycling. The water is then passed through an ion exchange system and either re-used or discharged to the sewer. Russo and Fischer [1989] discuss some of the costs of operating water recycling systems and show that a self-contained system operating thirty hours a week at 4 gal./min. can have water and energy costs as low as $1,000 per year, while an equivalent one-time-use system with treatment to bring the water up to local discharge standards can cost about $5,000 per year.

Of course, the real proof of a semi-aqueous system is its ability to clean. The EPA, DOD, and IPC have been running a test program for some time to independently verify the effectiveness of new cleaning solvents. In phase I, they established a baseline using the cleaner nitromethane-stabilized CFC-113/methanol azeotrope. They have since demonstrated that several solvents suitable for semi-aqueous systems, including Petroferm's Bioact^{R} EC-7 and DuPont's AXAREL^{TM} 38, are effective cleaners. This work has been confirmed in tests by several equipment manufacturers and users (see Table 5.3). For example, CalComp Corporation compared Bioact EC-7 with a CFC and found the ionic contamination, as measured by the micrograms of NaCl per square inch of printed circuit board, to be 20% less for the semi-aqueous cleaner (0.7 μg/sq. in. versus 0.875 μg/sq. in.). The equivalent cleaning performance of a standard CFC cleaner on the same RMA flux is 3.2 μg per sq. in. The equivalent cleaning performance of water on the same water-soluble flux is 6.9 μg per sq. in. Just as important as the better level of cleaning is the

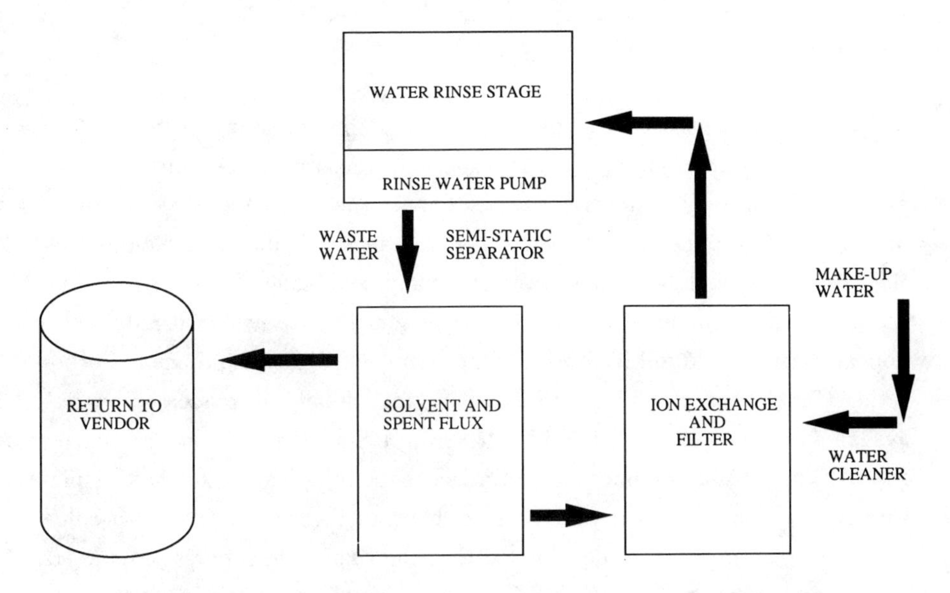

Figure 5.2: The principles of a closed-loop water rinse system

lower variability between printed circuit boards in the test for the semi-aqueous system. These and similar tests have demonstrated that the semi-aqueous system is an excellent alternative for cleaning SMT printed circuit boards using RMA and water-soluble fluxes.

Flux	Ionic Contamination (μg NaCl per sq.in.)
RMA	0.9
Water soluble	3.5

Table 5.3: Cleaning performance of BIOACT^{R} EC-7^{TM} proprietary semi-aqueous cleaner. (Bioact^{R} is a registered trademark of Petroferm^{TM}, Inc. EC-7^{TM} and Petroferm^{TM} are trademarks of Petroferm^{TM}, Inc.)

5.3.2 Aqueous systems

Manufacturers have used aqueous cleaning systems for many years to clean through-hole printed circuit boards, especially those using OA fluxes. However, the traditional systems could not handle the much smaller dimensions of the SMT components and their lower clearance heights when mounted on the printed circuit boards. Nevertheless, many companies persevered and developed ways to modify traditional systems for effective use with SMT components, down to 25-mil lead pitch. One company, Digital Equipment Corporation (DEC), has described its system in great detail and made the process available to others through ICOLP [Pickering 1990]. This section summarizes the DEC process — the microdroplet aqueous module cleaning process. A picture of an automatic cleaning machine is shown in Figure 5.3. DEC performed extensive experiments and determined that the significant process parameters are the water droplet size and the angle of impingement of the droplet spray on the printed circuit board. The basic principle of operation is the delivery of microdroplets of water and saponifier solution through a spiral hollow cone nozzle with a spray angle of 50 degrees. The nozzle is mounted at the end of a rotating arm, so the impingement angle of the droplets onto the board continuously varies. The saponifier softens the contaminants, making them easier to remove. Each printed circuit board is moved through the system on a conveyor belt. There are four stages: pre-wash, microdroplet wash, rinse, and drying. DEC engineers have described the nozzle as the critical component, even though it is not complicated. The equipment used by DEC in their first installation is a Stoelting CBW224 aqueous cleaner, modified by the addition of Bete Fog Nozzle, Inc., nozzles. DEC uses the system with Kester^{R} 229D solder paste and Kester^{R} Bio-Kleen^{R} 5779 saponifier. DEC has not tried other cleaners and paste/saponifier combinations, but it is reasonable to assume that the principles of the process can be applied to other equipment and chemicals.

The nozzles are mounted 4 in. above the printed circuit board and produce a spray of droplets at 500 psi, with at least 55% of the droplets having diameters in the range of 6 to 12 microns. The saponifier solution is maintained at 7% concentration and 160±5°F (71±2.5°C). The water is either de-ionized or softened to less than 2 ppm calcium and magnesium ions. Saponifier solutions are very alkaline — pH about 10 to 12 — and must be neutralized quickly to limit their tendency to attack the printed circuit boards. Also, operators must

Figure 5.3: Cleaning equipment

wear rubber gloves and eye shields at all times while using them. The conveyor belt is capable of speeds of up to 12 ft. per min., but the DEC system actually operates at 3 ft. per min. and can clean up to ninety printed circuit boards an hour.

The first cleaning stage which wets the printed circuit board and also cleans off particulate matter, operates at 10 to 15 psi. The second stage applies the microdroplets at a rate of 12 gal. per min. Because the nozzles are on a rotating arm, they actually spray the printed circuit board intermittently, allowing the aqueous solution to drain away between applications. Unlike conventional systems, which form a barrier layer of water around the components when they are flooded, intermittent spraying does not.

On exiting the wash stage, the printed circuit boards pass through neoprene curtains to remove excess solution. Next, they enter the rinse stage, where water cleans off the last of the cleaning solution. Finally, the printed circuit boards pass under air blowers and into a heated chamber, where they are dried.

As with the semi-aqueous systems, the detailed treatment of the waste water depends on local regulations. Most regulations cover: discharge temperature; effluent pH; the amount of heavy metals present; the biological and chemical oxygen demands (BOD and COD); solids in suspension in the effluent; the presence of grease, oils, and fats; and the total amount of toxic organics present (TTO). A self-contained waste-water system with continuous treatment of the rinse water and the capability for water make-up is often the best approach.

Other manufacturers have developed different approaches to aqueous cleaning. Hollis and Treiber both use fan-shaped jets of water to spray the printed circuit boards, which are passed below the jets. Hollis uses the trademark Hurricane Jet™ [Lowell and Sterritt 1990]. The unmodified Stoelting system uses many jets placed around the printed circuit boards so some water strikes them from all angles. No one seems to have published results on the relative effectiveness of the different techniques. Manufacturing engineers should be careful to review them all before making a final selection.

5.3.3 No-clean processes

One obvious way to improve cleaning is to eliminate it entirely. Great progress has been made recently in developing and applying no-clean fluxes (NCFs), and several are now available commercially. The major characteristic of an NCF is the low solids content. Other types of flux have a total of 25 to 35% solids in a solvent that facilitates application of the flux; low-solid fluxes have 1 to 5% solids. There are two constituents in the solids: the activator, which creates the wettable surface for the solder, and the vehicle, which acts as a high-temperature solvent, facilitating the final distribution of the flux over the surfaces to be soldered. In most fluxes, the activator is the smallest constituent, but in low-solid fluxes, the vehicle is often the smallest constituent.

Companies adopting NCFs find that they have to change the whole process. AT&T has published a lot of information on the work they have done in this area [Parikh 1991]. They emphasize the importance of careful application of flux. Controlled amounts of the flux are placed in the required areas. At first, an ultrasonic nozzle from Sona-Tek Corporation was used to apply the flux, but an airless spray gun that produced a long, flat spray, parallel to the direction of the printed circuit board later replaced it. This spray is moved backwards and forwards across the width of the printed circuit board. The current system can accommodate printed circuit boards which are up to 24 in. across. It operates with ±15% uniformity on one printed circuit board, ±5% repeatability from printed circuit board to printed circuit board in a batch, and ±10% long-term repeatability. Any overspray is captured and recycled. The flux reservoir is closed, reducing flux testing and alcohol make-up and virtually eliminating the disposal of spent flux. AT&T has stated that using an NCF in a closed system in one production line saved about $100,000 per year in solvents and about $20,000 per year in flux testing and solvent make-up costs.

NCFs are usually soldered in an inert atmosphere of nitrogen, with less than 500 ppm of oxygen. This process requires considerable changes to the mass reflow solder equipment, but it offers a more stable process with a wider window than reflowing solder in air. NCFs do leave some residue on the printed circuit board; it may not be a source of corrosion, but it degrades the cosmetic appearance of the printed circuit board. Using a bleached flux, which tends to leave a colorless residue, is a partial solution to the problem.

Using NCFs only addresses part of the cleaning problem — the elimination of contamination from the flux residue. Manufacturers must improve their processes to eliminate contamination from workers and other sources, such as laminates, adhesives, solder, and the environment. One common source is the creation of solder balls as water vapor, which is trapped between the laminates of a printed circuit board and evaporates during the high-temperature phase of the soldering process, bubbles through the solder and splatters it. Unless problems like this are addressed during the process specification, cleaning the printed circuit boards will still be necessary, destroying any advantage of using NCFs. The focus of these process improvements should be on establishing a clean, room-like environment and improving worker awareness and training. There is no need to go to the extreme of conducting a semiconductor facility, but serious thought should be given to going to a class 10,000 environment for printed circuit board assembly operations.

As more manufacturers move to very-fine-pitch SMT components, more will recognize the benefits of NCF, because it eliminates the need for cleaning and its associated handling. This result is advantageous because of the fragile leads on these devices. The associated process improvements outlined in the previous paragraph will be very helpful in establishing good yields for assembling very-fine pitch SMT printed circuit boards.

An alternative NCF technique is to use a reactive gas as the flux during the soldering process [Marczi 1990], but this technique does not eliminate the need for some form of flux that is activated by the reactive gas. More work is needed on this approach before many manufacturers will adopt it. There are concerns about the safety of the gases used, and more work must be done to establish the economics of the process. Nevertheless, it may be the only satisfactory approach for very large SMT components with lead pitches of less than 12 mils.

5.4 Monitoring cleanliness

Another challenge is to measure printed circuit board cleanliness. Most measurement techniques are lengthy and not suited to in-line process control. The surface insulation resistance (SIR) test is a direct measure of the cleanliness of the PCB surface. The procedure starts with defining an interleaved pattern of conductors on the surface of the printed circuit board, as shown in Figure 5.4. After the cleaning process is completed, a voltage is applied between the two electrodes and the leakage current is monitored. If the process has left ionic residue on the surface of the printed circuit board, there will be a small current flow across the surface. If the voltage is applied for some time and the contaminants are creating corrosion, the leakage current will also increase with time. However, the test is time-consuming and its results can be obscured by external factors, such as leakage currents through the PCB. The electrode pattern also takes up space on the printed circuit board. Nevertheless, it is a commonly used cleanliness test whose results can be directly correlated to printed circuit board reliability.

Other cleanliness tests attempt to dislodge any residues from the PCB surface and then measure the amount in the fluid used. MIL-P-28809 specifies a 75% isopropanol and 25% de-ionized water solution, by volume. While the

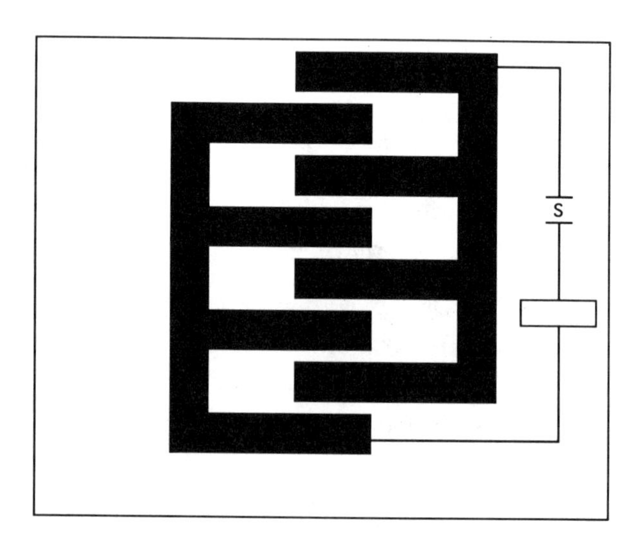

Figure 5.4: An electrode pattern for surface insulation resistance testing

measurement of the residue entrained in the fluid may be very accurate, there is often little certainty that all the residue was removed. Omegameters (manufactured by Kinco Industries), Ionographs (Alpha Metals), or Contaminometers (Protonique) are commonly used to check the amount of residue present; they are generally regarded as giving a good indication of the relative cleanliness of printed circuit boards in a batch but none of the techniques really gives an absolute measure.

IPC test method IPC-TM-650 describes ways to measure non-ionic contamination though these tests are not usually as important as the tests for ionic contamination. The IPC methods do not identify the contaminant or separate mixtures into their constituent parts.

5.5 References

Agnew, Jeremy. Choose Cleaning Solvents Carefully. *Electronic Design* 5, (March 1, 1974); 54- 57.

Capillo, Carmen. *Surface Mount Technology.* New York: McGraw-Hill, 1990.

Hayes, Michael E. Semi-Aqueous Defluxing Using Closed-Loop Processes. Paper IPC-TP-898: Second International Conference on Flux Technology, 1990.

Lowell, Charles R., and Sterritt, Janet R. An Aqueous Cleaning Alternative to CFCs for Rosin Flux Removal. Paper IPC-TP-893: Second International Conference on Flux Technology, 1990.

Marczi, M., Bandyopadhay, N., and Adams, S. No-Clean, No-Residue Soldering Process. *Circuits Manufacturing* 30 (February 1990) 42-46.

Murray, Jerry. Ultrasonic Cleaning: What's the Buzz ? *Circuits Manufacturing* 30 (January 1990) 72-74.

Parikh, Girish. AT&T Achieving Environmental and Quality Goals. *Circuits Assembly* 2, (December 1991) 60-62.

Pickering, Ray. Digital's Microdroplet Aqueous Cleaning Project. Paper IPC-

TP-82: Second International Conference on Flux Technology, 1990.

Digital Equipment Corporation. Aqueous Microdroplet Module Cleaning Process. ICOLP, 1990.

Prasad, Ray P. *Surface Mount Technology.* New York: Van Nostrand Reinhold, (1989), 471-513,

Russo, John F., and Fischer, Martin S. Recycling Boosts Aqueous Cleaners. *Circuits Manufacturing* 29 (July 1989) 34-39.

Chapter 6

Reliability and Quality

Shailendra Verma
CALCE Electronic Packaging Research Center
University of Maryland
College Park, MD

Reliability and quality of solder joints are topics usually included under one heading, though they address different issues. Reliability, as will be explained, is the ability of the solder joint to meet design life or usage criteria. Quality assurance, on the other hand, is an ongoing task associated with the monitoring of materials and process parameters to ensure that previously qualified process and design parameters continue to be maintained with the specified tolerances. The importance of both aspects can hardly be over-emphasized when the United States produces, on the average, three hundred million solder joints annually.

Lead-tin alloys have been used in electrical connections without many major problems since the advent of electricity. However, with the introduction of surface-mount technology, solder is being used not only as an electrical connector, but also as a medium to provide mechanical support. Increasing concern is also being experessed with lead and chlorofluorocarbons in the environment. Increasing density, higher operating temperatures, and increased competition to develop highly reliable products have also been causes of concern.

The solder joint is a critical hardware link in an electronic circuit; the failure of the joint can disrupt the functionality of the entire device. Many expensive mishaps have been attributed to solder-joint failure. For example, the Hubble Space Telescope project was reported to have experienced several solder-joint failures of lap-soldered flat packs during normal operation [Goddard Space Flight Center Report]. On the other hand, a failure in a common consumer product can cause customer inconvenience, warranty liability, and a reduction in product demand. If even a fraction of failures were caused by unreliable joints, the cumulative loss in terms of money is extensive. Regardless of product application, reliability of solder joints is vital and is a subject of grave concern in the modern electronic equipment industry.

This chapter introduces the concept of reliability and quality, and highlights major issues concerning reliability and quality of solder joints. An overview of the variables that affect the life (reliability) and yield (quality) of solder joints is presented as are testing and verification methods and procedures. Reliability is defined as the probability of a product performing its intended function without failure under stated conditions. The "intended function" of the device has to be defined to assess whether performance is adequate. Moreover, the intended function should be defined "for a given period" or designed life that serves as a time frame for reliability calculations. Finally, the operating conditions ("stated conditions") need to be known to develop specific reliability requirements. A given reliability requirement for solder joints for an F-16 aircraft will demand far more stringent testing and process control than for a home computer.

Reliability is measured by three parameters: failure or hazard rate, mean life cycles to failure, and probability of survival. Definitions and basic concepts of these parameters can be found in any reliability text. The most critical issue for the reliability engineer is to quantify these parameters in the relevant environment. There are two common approaches to calculating the reliability function of a component or device. The first is based on statistical empirical models fit to field data of hardware failures. The other is based on the physics or mechanics of the relevant actual failure mechanisms, the accompanying material properties, and stress analysis. The approach is generic and applicable to both existing and future devices.

This chapter treats the reliability of solder joints from a physics-of-failure perspective. The emphasis is on understanding the mechanisms of failure and developing models to help predict them. The approach is not new, but has gathered importance over the last few years due to the inadequacies of the military standards and handbooks. Some progress has been made in modeling solder-joint failure, but it is a very complicated subject and many fundamental issues including creep, viscoelasticity and fatigue still need to be clarified before implementing this approach. A combination of physics and empirical analysis is required to solve problems temporarily, while research on the basic issues continue.

6.1 Reliability concepts

The reliability function can be depicted by plotting the hazard rate *f(t)* along the Y-coordinate and the operating time or life, t, on the X-coordinate (see Figure 6.1). This curve is also called the "bathtub" curve, because of its profile (curve 1, Figure 6.1).

Generally, the curve can be separated into three distinct regions. Each region represents failures in the electronic device due to different factors. Region I depicts decreasing rates of failure very early in the life of the device. The early failures are sometimes called "infant mortality" and represent failures in the field due to mishandling, poor assembly, or manufacturing defects. For example, a connection with inadequate solder may pass the electrical testing but fail early in the field. This region comprises early wear-out failures that are, ideally, negligible in well-designed and manufactured devices. Region I can be eliminated by effective screening.

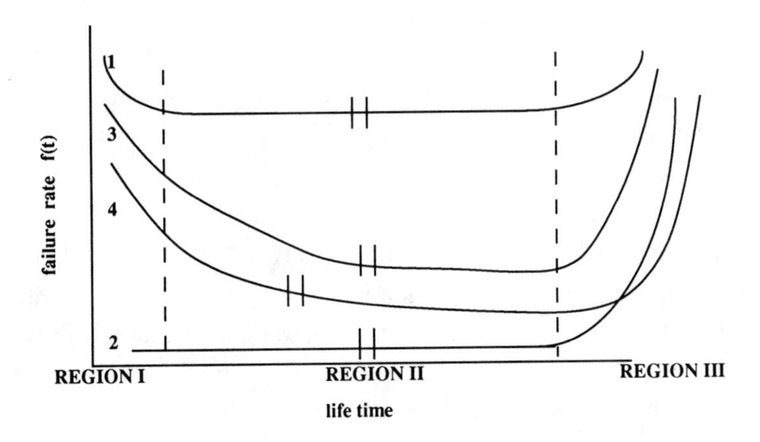

Figure 6.1: Curves 1 to 4 show the different possible failure distributions. Curve 1 shows a typical bathtub failure distribution curve, while curve 2 represents an ideal situation in which the early and design failure rates are constant and low. Curves 3 and 4 depict possible variations from the ideal situation.

Region II describes the operating rate of failure of the device. A well-designed solder joint has a very low hazard rate of nearly zero (as shown by curve 2 in Figure 6.1).

Region III represents the stage when the solder joint begins to wear out and shows an increasing rate of failure. The design team's task is to design and manufacture a product which does not begin to enter Region III until the tested or expected life of the product is achived. However, while early wear-out typically results from poor design, late wear-outs imply over-design.

The hazard rate distribution is often unpredictable in real life, because failures cannot be so distinctly classified and because solder joints have many ways to fail. The real-life scenario is far more complex than a simple bathtub curve; and Curves 3 and 4 of Figure 6.1 are examples.

The rate of failure depends greatly on the material, variabilities in manufacturing and assembly processes, design life, and load. Failures occur when the stress or combination of stresses experienced by the solder joint exceed its strength.

The objective of the manufacturer should be to understand the failure pattern — the wear-out and early failure mechanisms — by using predictive and experimental methods. These require investigation of the mechanisms that cause failure and should lead to steps to prevent its occurrence through design modification, process control and, when necessary, screening. Estimation of reliability is important to evaluate the testing and screening efforts required to manufacture the solder joint, the status of production, and the defects being introduced into the manufacturing or design process, and as a guarantee for the customer.

The methodology to achieve high reliability has frequently changed over the years, with changes in technology and customer demands. The present thrust to achieve high reliability focuses not only on sound design, but also on good manufacturing practices and rigorous process control. Process control also comes under the purview of quality assurance, so reliability and manufacturing departments need to work as a team. The industry is trying to move away from expensive testing and inspection towards building reliability into the products.

The prime goals of any reliability program are to validate nominal design and manufacturing parameters, and to specify nominal values and allowable tolerances on design variables, such as geometry and material properties, and on manufacturing variables, such as process flow and machine setting. Validation involves justification of design and manufacturing variables so as to maximize the reliability of the solder joint while minimizing cost. The process requires relating the requirements which also serves as the basis for specifying tolerance and defining a defective product.

Validation encompasses simulation and testing methods required to establish the correctness of the nominal values and allowable tolerances specified for design and process variables. In order to successfully qualify a solder joint or mounting technology, it is important to know the design and manufacturing process, the loads throughout the life-cycle, the associated potential failure mechanisms, and the cost penalties associated with each design and manufacturing decision. Physics-of-failure models are important validation tools involving study of the physics of the solder joint to predict its life. The appropriate failure mechanisms — the physical processes by which stresses damage solder joints — must also be identified. Failure mechanisms, which relate the material, the joint geometry, and the load to the life of a solder joint, are presented in Section 6.2; the relationship of process parameters to failures is dealt with in Section 6.5.

Validation requires acclerated reliability testing, which is discussed in Section 6.3. Such testing should to be performed only during initial product development, or during design or manufacturing changes. Once the solder joint is qualified, routine lot-to-lot requalification is redundant and an unnecessary cost item.

6.2 Wear-out failure mechanisms

Every element in nature has a natural degradation process that, with time and usage, weakens its ability to function and eventually results in its failure. Solder degradation and failure are caused by field loads, such as thermal fatigue, vibration, corrosion, and mechanical shock. Thermal fatigue is caused by unequal thermal expansions of the substrate and component during temperature cycling. Cyclic deformation can also be introduced by vibrational loading or a combination of both thermal cycling and vibrational loading, as in aircraft or automotive applications. While thermal variation frequencies usually range from five cycles per minute in telecommunication equipment to one cycle per day in computer processors, vibration frequencies typically range from 10 Hz to 2,000 kHz, which can translate to over 100,000 cycles a day [Barker 1992].

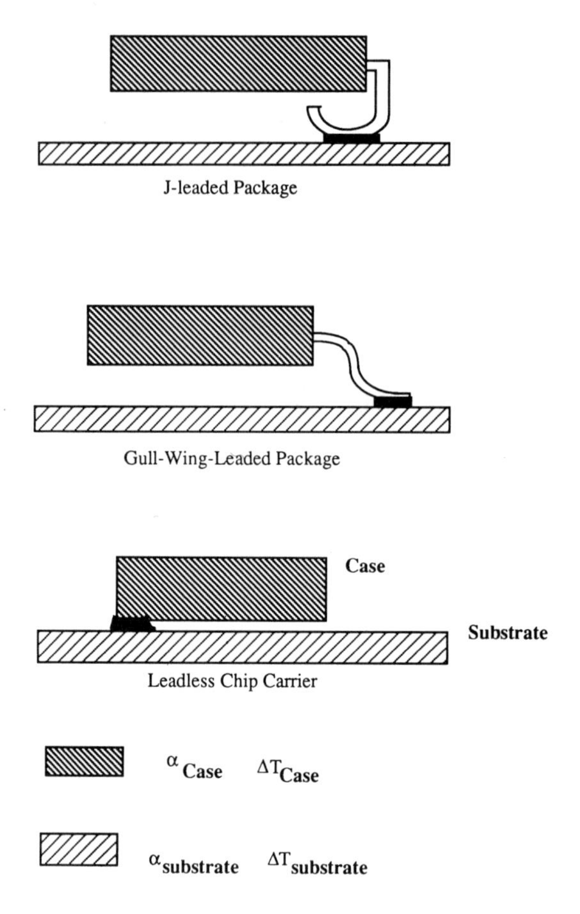

Figure 6.2: Common surface-mount configurations

Studies indicate that about 80% of solder and lead failures are attributed to thermal cycling and the rest to vibrations [Sandor 1990].

The complex problem of solder-fatigue life prediction is influenced by many variables. Solder fatigue presents a unique case in which time-dependent failure mechanisms, such as creep, viscoelasticity, and viscoplasticity, contribute to degradation. Research on many of the phenomena is still in the very basic stages.

Some surface-mount solder-joint configurations are shown in Figure 6.2. When a component undergoes a change in temperature (ΔT), the case of the component expands by $\alpha_{\text{case}}\Delta T_{\text{case}}$; the substrate, to which the solder is attached, expands by $\alpha_{\text{substrate}}\Delta T_{\text{substrate}}$, where ΔT_{case} represents the temperature swing for the case, $\Delta T_{\text{substrate}}$ represents the temperature swing for the substrate, α_{case} represents the coefficient of thermal expansion of the case, and $\alpha_{\text{substrate}}$ represents the coefficient of thermal expansion of the substrate. The resulting mismatch is given by:

$$\Delta(\alpha\Delta T) = |\alpha_{\text{substrate}}\Delta T_{\text{substrate}} - \alpha_{\text{case}}\Delta T_{\text{case}}| \tag{6.1}$$

The situation looks deceptively simple in the above equation. However, typically when a component is powered, a temperature gradient results through the component in various ways dependent on the heat transfer paths. The temperature gradient over the substrate will also cause bending and bowing of the substrate, resulting in compressive and tensile strains in the solder joint. The state of strain in the solder joint is not simple uniaxial shear as assumed in the above equation, but multiaxial. Furthermore, there are mismatches between the co-efficients of thermal expansion (CTE) of the solder material, the substrate, and the lead material, which cause local strain in the solder. Equation 6.1 accounts only for the CTE mismatch of the case material and the board, neglecting the local effect. The causes of strain are depicted in Figure 6.3 [Hall 1991].

When a device is switched on, the component is heated, which increases the operating temperature of the solder joint. The temperatures usually are beyond $0.5T_m$ (T_m is the melting point of solder). At such high temperatures, the mismatch deformation comprises not just elastic deformation processes, but also plastic and creep strain:

$$\varepsilon_{\text{total}} = \varepsilon_{\text{elastic}} + \varepsilon_{\text{plastic}} + \varepsilon_{\text{creep}} \tag{6.2}$$

where ε is the strain (deformation per unit).

When a load or stress is applied to a metallic material, such as solder, it undergoes deformation. If the stress is below the yield point, the material responds, for most part linearly and reversibly, with small lattice deformations called elastic strain. For stresses beyond the yield point, the material undergoes a temperature-dependent permanent deformation called plastic strain. Plastic strain is caused by the motion of crystal defects in the material due to higher loads. Unlike elastic strain the strain is irreversible.

As the plastic deformation increases, the dislocation chains pile up and the plastic flow stress increase; the increase in dislocation requires an increas-

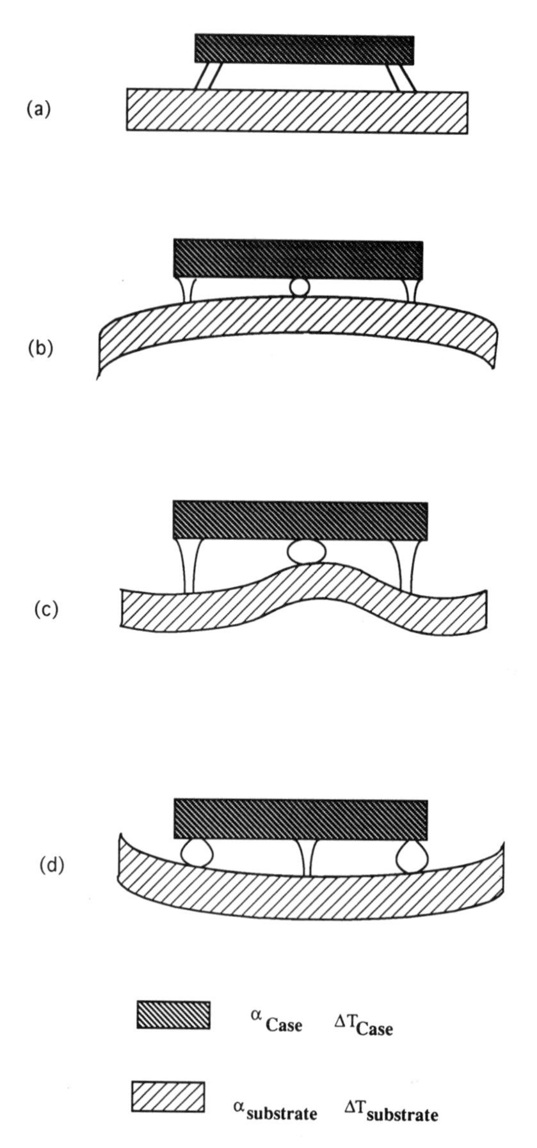

Figure 6.3: A leadless solder joint experiences multiaxial strain depending on the deformations of the substrate.

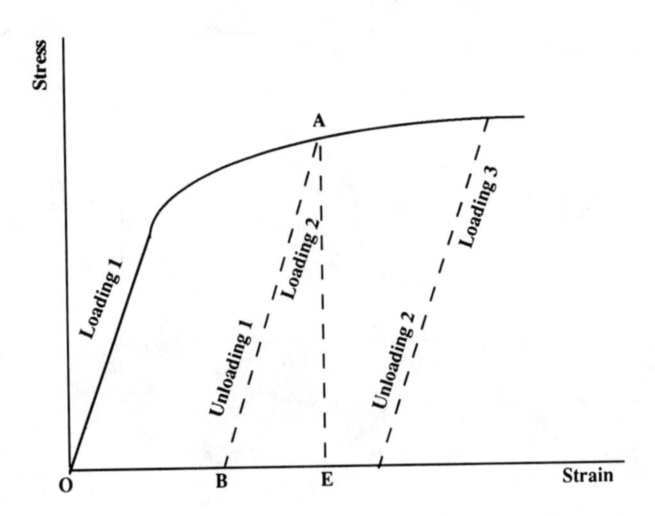

Figure 6.4: The loading and unloading sequence and shows the effect of strain hardening.

ing stress for continued deformation beyond the yield point. This phenomenon, called strain-hardening, is observed in solder, as shown in the stress-strain diagram (Figure 6.4). Another important effect of strain hardening occurs during cyclic loading of solder and can be explained with the help of Figure 6.4. When the solder is unloaded from point A on the stress-strain diagram, the permanent deformation is represented by OB, while the recoverable elastic deformation is BE. If loaded again, the solder follows path BA until it yields at point A. The figure indicates the increase in yield strength for the second loading as a result of the first loading into the plastic range [Shames and Cozzarelli 1992].

When a constant load is maintained over a long period of time, some materials exhibit a continuous increase in strain. This phenomenon, particularly prominent when the material is at a high homologous temperature (as is solder), is called creep. The cause is attributed to increasing dislocations, diffusion of vacancies, and sliding of grains past each other. (A typical creep curve for solder is shown in Figure 6.5.) Creep strain is strongly dependent on the load, the temperature at which the load is applied, and the duration of the load application, as depicted in Figure 6.6 [Tribula 1989]. High-temperature creep allows stress relaxation and minimizes stress concentration points in solder, from which cracks can form and grow. On the other hand, large deformations

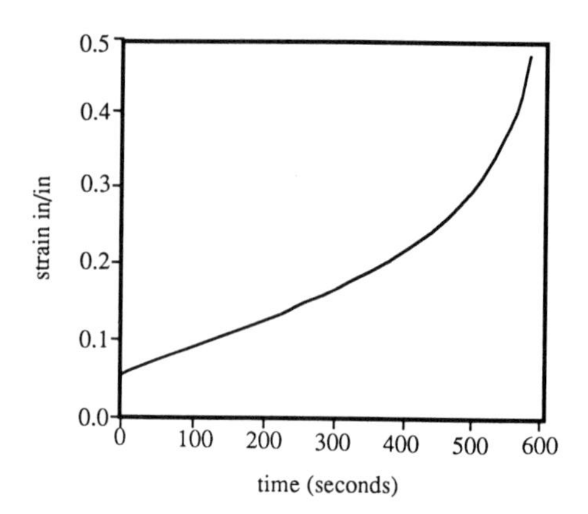

Figure 6.5: A typical creep curve for a 58Sn-40Pb-2Sb alloy

can result in microstructure degradation in alloys like solder, which depend on the phase morphology for creep resistance. Recrystallization also occurs as dislocations entangle and grow [Tien et al. 1991]. Thus, in solder joints, high strain rates can be both benign (they prevent stress concentration) and hazardous to reliability.

Most solder joints fail due to the accumulation of creep and fatigue during repeated load application. The fatigue life of metallic materials is usually depicted as the number of cycles to failure plotted against the strain range, as shown in Figure 6.7. Fatigue failure takes place in stages: crack initiation, slip-band microscale crack growth, macroscale crack growth on planes of high tensile stress, and ultimate ductile failure [Lau and Rice 1985]. The time-dependent creep deformation in solder makes the temperature history an important variable.

The adverse effect of creep on fatigue can best be understood by the reference to stress-strain plot shown in Figure 6.8. Curve A shows a cyclic strain hysteresis loop of a material stress relaxation, while curve B shows the material without stress relaxation. The damage occurring in a fatigue cycle is proportional to the area within the cyclic strain hysteresis loop. This hysteresis represents the inelastic work done on the solder; such work decreases the solder's ductility and, thus, its fatigue life. The increase in area signifies

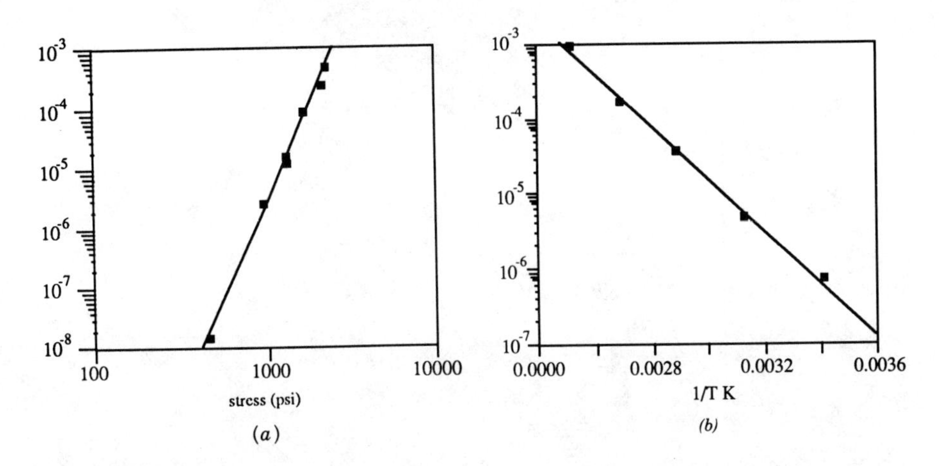

Figure 6.6: Plots showing the dependence of creep on (a) stress and (b) temperature.

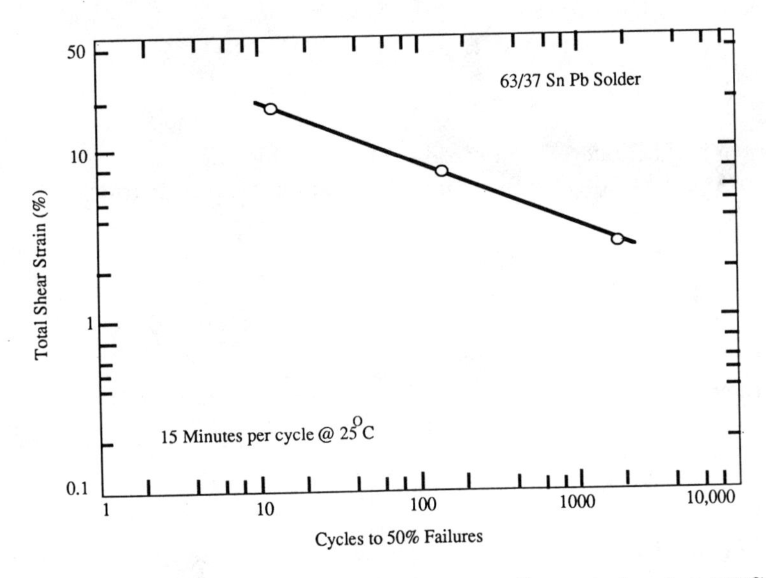

Figure 6.7: Solder joint fatigue life versus cyclic strain level at 25°C

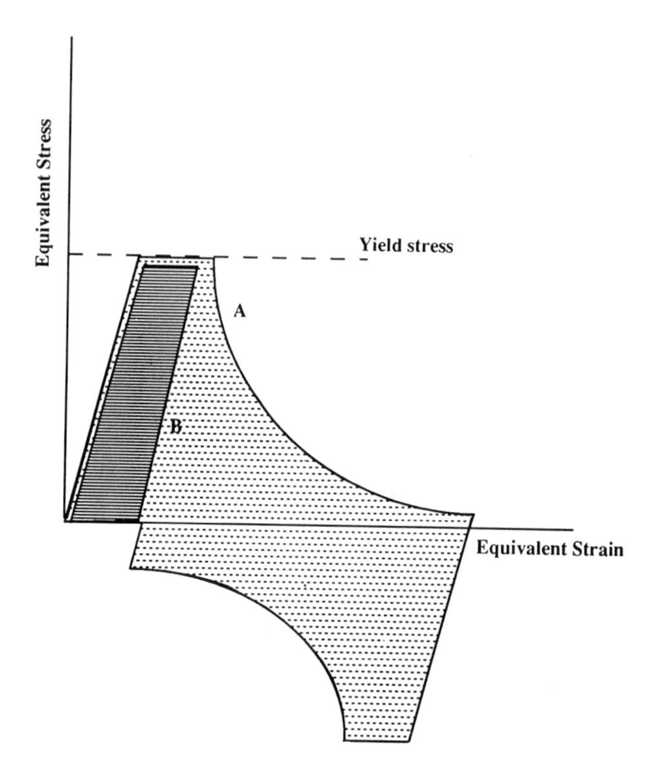

Figure 6.8: The difference in the area of the loading and unloading sequence of curves A and B indicates the damage caused by stress relaxation

the additional damage incurred due to stress relaxation. Some researchers have also found evidence of microscopic creep-fatigue interactions. Fatigue striations and river lines occurred along with intergranular cracking and surface voids in thermally cycled solder joints, indicating both creep and fatigue. Creep voids increase with the increase in the number of cycles experienced by the solder [Attarwalla et al. 1991].

The problem is to find a comprehensive life-prediction technique that considers the fatigue-creep interaction in solder joints and estimates life in terms of the number of load cycles. The important variables in the problem include the following:

- cyclic temperature range;
- mean temperature;

- dwell times;
- dwell temperatures;
- loading and unloading rates;
- lead and solder joint geometry;
- solder microstructure;
- the presence of intermetallics;
- the presence of alloying and trace elements; and
- soldering defects [Oyan 1991].

However, it has not been possible to incorporate all the variables in solder-joint life prediction; only the first six are typically used explicitly in modeling wear-out failure.

Existing life-prediction methodologies generally use three approaches: strain-based, stress-based, or energy-based.

6.2.1 Strain-based approach

A strain-based approach is used when the stress state of the solder joint depends on the deformation applied to it, as in the case of thermal loading. Thermal fatigue failures take place in less than 10^4 cycles, and are classified as low-cycle fatigue. The most popular method used to predict fatigue life, N_f, is the Coffin-Manson relation [Morrow 1965]. In a generalized form it can be written as:

$$\frac{\Delta\varepsilon}{2} = (\frac{\sigma_f}{E})(2N_f)^b + \varepsilon_f(2N_f)^c \quad (6.3)$$

where $\frac{\Delta\varepsilon}{2}$ = applied total strain amplitude , ε_f = fatigue ductility coefficient, N_f = number of cycles to failure, c = Coffin Manson fatigue ductility exponent, σ_f = fatigue strength coefficient, E = modulus of elasticity, and b = Basquin fatigue strength exponent.

This equation comprises an elastic part and a plastic part. The plastic curve dominates the low-cycle fatigue; often, researchers simplify the equation by considering only the plastic component. However, the limitations of the Coffin-Manson equation are that it does not consider fatigue-creep interaction and is valid for completely reversed, constant amplitude, sinusoidal loadings. Coffin therefore suggested a modified form that takes into account factors affecting creep, such as temperature, dwell time, and cycle time [Coffin 1973]:

$$\Delta\epsilon = C_2[N_f \nu^{k-1}]^{-\beta} + \frac{AC_2^n}{E} N_f^{-\beta n} \nu^{-\beta n(k-1)+k_1} \tag{6.4}$$

where C_2, A, β, n, k, and k_1 are temperature-dependent material fatigue properties and ν is the frequency of loading.

Engelmaier [1985] has also proposed an approximate model along the lines of a Coffin-Manson relationship by assuming that plastic strain is pure shear. The equation elegantly accounts for the effect of temperature and dwell time on creep-fatigue behavior using empirically derived relations for 60Sn-40Pb solder. The model can be expressed as:

$$N_f = \frac{1}{2}\left(\frac{\Delta\gamma}{2\varepsilon_{f'}}\right)^{\frac{1}{c}} \tag{6.5}$$

where

$$2\varepsilon_{f'} = 0.65 \tag{6.6}$$

$$c = -0.442 - 6\times10^{-4} T_{SJ} + 1.74\times10^{-2} \ln\left(1 + \frac{360}{t_D}\right) \tag{6.7}$$

and where T_S, T_C = steady-state operating temperature for substrate and component, respectively ($T_S > T_C$ for power dissipation in component); $T_{SJ} = 0.25\times(T_S + T_C + 2T_o)$, representing cyclic solder temperature; T_o = temperature during half cycle.; t_D = half-cycle dwell time (min), average time available for stress relaxing at T_S, T_C and T_o; $\Delta\gamma$ = cyclic total plastic shear strain range in solder after complete stress relaxation.; and $\varepsilon_{f'}$ = fatigue ductility coefficient (defined for 60/40 solder above).

Engelmaier's model is simple and effective, but the empirical coefficients exist only for eutectic solder. The model is valid for simple shear strain but the actual solder joint is subjected to multiaxial strain.

The "strain-range partitioning method" is another means proposed to account for creep-fatigue interaction in predicting solder life [Halford et al. 1979]. The method involves recording the plastic and creep strains separately from a uniaxial completely reversed loading. Four partitionings are possible: pp (plastic in tension and compression), cp (creep in tension and plastic in compression), pc (plastic in tension and creep in compression), and cc (creep in tension and compression). The components are then related to cyclic life by a Coffin-Manson-type equation. (This partitioning is shown in Figure 6.9.) The number of cycles to failure can be predicted by summing up the damage due to each component:

$$\frac{1}{N_f} = \frac{F_{pp}}{N_{pp}} + \frac{F_{cc}}{N_{cc}} + \frac{F_{cp}}{N_{cp}} + \frac{F_{pc}}{N_{pc}} \quad (6.8)$$

where N_f = predicted cycles to failure and N_{pp} = cycles to failure for a cycle of type pp, while F_{pp} is the fraction of the inelastic strain range of the type pp (similar notations are used for the cc, pc and cp cycle types).

Strain-range partitioning has its limitations; it requires extensive data, which is also difficult to measure, to be used correctly. This technique was proposed and tested for steel, but there is insufficient evidence that it can be applied effectively for solder joints [Sandor 1990].

These prediction methodologies are confined to predicting solder life in low-cycle fatigue. The variables accounted for are cyclic temperature range, mean temperature, dwell times, dwell temperature, loading-unloading rate, and, to an extent, lead geometry. However, in pure vibrational loading, the above parameters cease to play a critical role and a stress-based approach must be employed.

6.2.2 Stress-based approach

This approach is used to model high-cycle fatigue failures, which are typically caused by shock and vibration; these failures are stress-dominated or elastic-strain-dominated events and occur after 10^4 cycles. The loads are generally low and the elastic strains are much greater than the plastic ones.

Stresses or loads are induced in the solder joint and the substrate by dynamic displacements caused by vibrations. The stresses are proportional to the square of the natural frequency, a critical parameter in calculating fatigue damage. Natural frequencies can be approximated from simple plate equations; finite element techniques are used for more accurate calculations, by modeling the printed circuit board with components placed on it and applying realistic boundary conditions.

Once the natural frequency is known, two approaches can calculate the life of the solder joints. The first, developed by Steinberg [1988], uses an empirical equation for the maximum allowable displacement, d, at the center of the board:

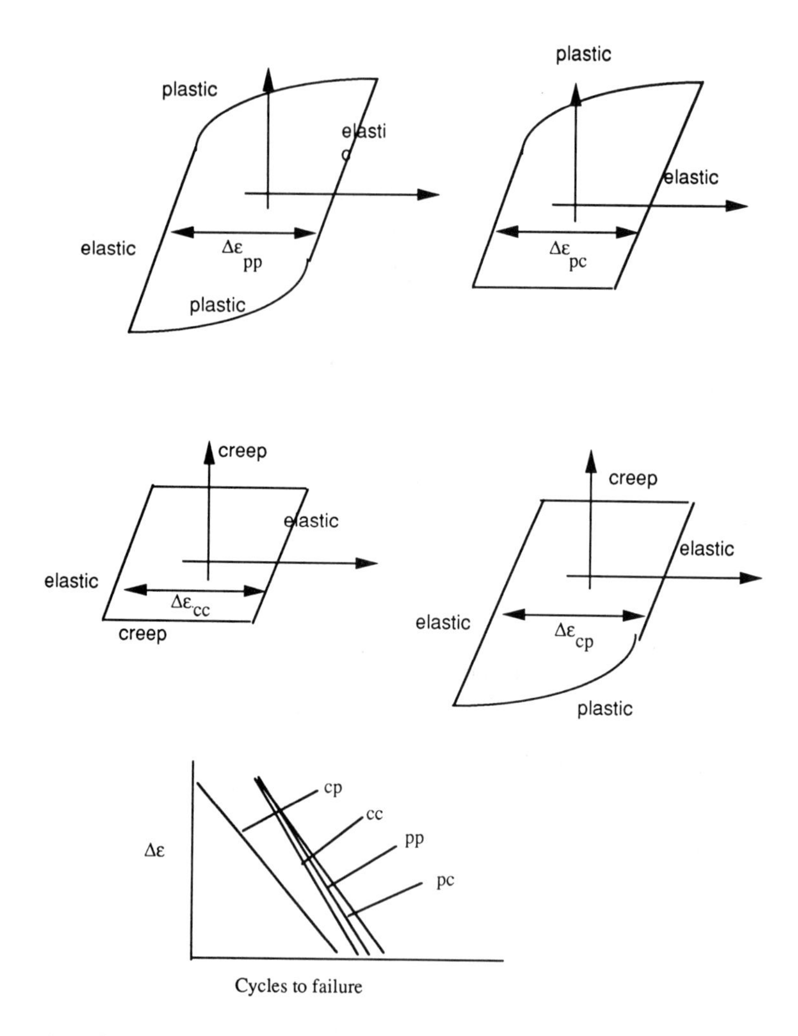

Figure 6.9: Hysteresis loops of the four basic types of inelastic strain-range, and a schematic strain range-life relationship

$$d = 0.00022 \frac{B}{ct\sqrt{L}} \tag{6.9}$$

where B = the length of the printed circuit board edge parallel to the component located at the center of the board (in inches); L = length of component (in inches); t = thickness of printed circuit board (in inches); and c = 1.0 for standard DIP, 1.25 for a DIP with side brazed leads, 1.0 for pin-grid array with four row of pins, and 2.25 for LCC.

One of the underlying assumptions in Equation 6.9 is that the printed circuit board is supported on all four sides. Steinberg rates this equation for ten million stress reversals under harmonic vibrations (sinusoidal) and twenty million stress reversals under random vibrations.

Fatigue life can be more accurately calculated by analyzing the stresses in the solder joint due to the vibrational loads. The relative displacements between the component and the board, in terms of the radius of curvature, are obtained from finite element analysis; the out-of-plane lead-attachment stiffness, K_y, is then used to find the force transmitted across the solder joint. The deformation of the component case is assumed to be negligible with respect to deformation of the board, and the deflection in the solder joint can be determined. The average maximum stress amplitude in each solder joint is calculated by dividing the force by the nominal cross-sectional area of the solder, while accounting for the stress concentration factor. Basquin's high-cycle fatigue relation is then used to calculate the fatigue life:

$$\sigma N_f^b = \text{constant} \tag{6.10}$$

where N_f = predicted number of operating cycles before failure and σ = normal stress in the solder joint. Basquin's high-cycle fatigue life considers only the elastic component because the plastic component is small enough to be neglected.

High-cycle fatigue analysis can also be analyzed through a fracture mechanics approach. The fatigue crack growth rate, $\frac{da}{dN}$, is related to the stress intensity factor range, ΔK, by the following relation:

$$\frac{da}{dN} = A(\Delta K)^p \tag{6.11}$$

The constants A and p vary in different regions of the stress-intensity-factor range and the fatigue-crack growth range. This is Paris' power law for crack propagation under completely reversed uniaxial loading in isotropic materials. ΔK is related to the applied stress level and to the instantaneous crack size by the following relation:

$$\Delta K = Y \Delta\sigma \sqrt{\pi a} \tag{6.12}$$

where Y is a calibration factor dependent on specimen geometry, $\Delta\sigma$ is the amplitude of the applied stress level, and a is the crack size. The fracture mechanics approach is not effective for low-cycle fatigue analysis, because the concept of the stress-intensity factor is restricted to linear elastic situations and does not apply when there is large-scale plasticity at the tip. The J-integral approach is used for plasticity at the crack tip, but there are serious limitations in the approach. These include the common occurrence of yielding at the crack tip, the instability of the material, and the problem of measuring the crack length in small solder joints [Sandor 1990].

6.2.3 Energy-based approach

A method based on energy partitioning has been proposed by Dasgupta et al. [1991, 1992]; here, the stress-strain history has been partitioned to obtain the stored elastic-strain energy, the plastic work dissipation (Wp), and the creep work dissipation (Wc), per cycle. The elastic-, plastic-, and creep-partitioned work is then related to the Manson-Coffin power law relationship to calculate the number of cycles to failure due to the energy dissipated. The total life cycles to failure are calculated by linearly summing the damage, as in strain-range partitioning.

This theory was implemented using the finite element method [Dasgupta et al. 1992a] on a surface-mount J-leaded device, as shown in Figure 6.10. The contour plot of the life of the solder joint generated from energy-partitioning analysis is shown in Figure 6.11. The figure shows that a crack would begin and propagate along the length of the J-lead after approximately 5,000 cycles for the simulated component.

6.3 Reliability testing

Testing solder joints for their wear-out life in normal operating conditions is expensive, time-consuming and often not feasible. Failure mechanisms are inves-

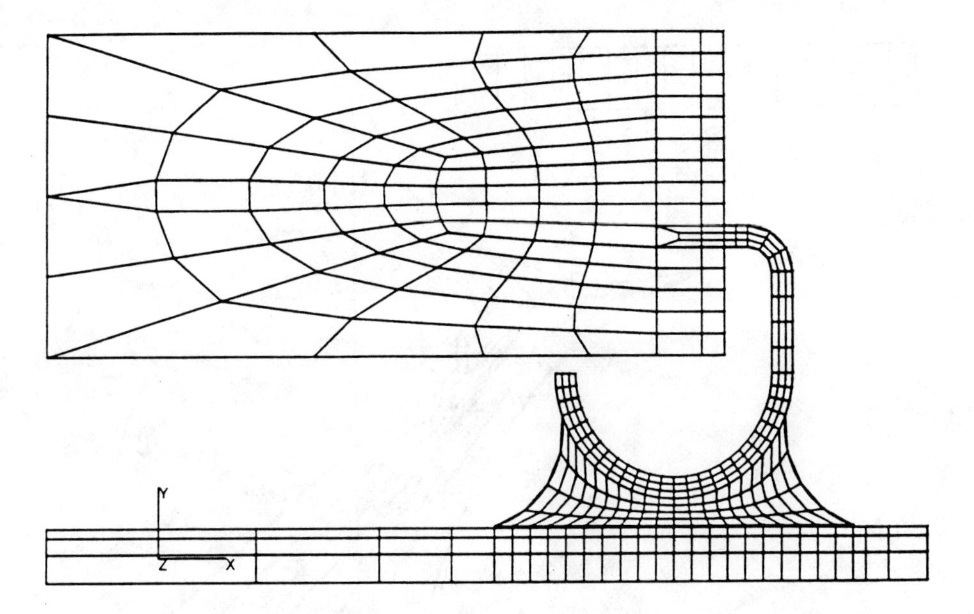

Figure 6.10: Finite element model of a surface-mount J-leaded device

tigated and reliability is measured by accelerated testing or accelerated-stress life testing.

The goal of accelerated testing is to accelerate the time-dependent wear-out failure mechanisms and the damage accumulation to reduce the time to failure. This is achieved by applying loads or stresses that are more severe than the field operating level. The correlation of the stress level to the time taken for testing can be explained as shown in Figure 6.12, which shows a typical fatigue life of any material. When the stress range is at normal field conditions (Level 1), the life predicted is N_1. For testing purposes, if the stress range were to be increased to Level 2, the test life would be N_2. The actual life could be calculated by:

$$N_1 = N_2 \times \text{Level}\,2/\text{Level}\,1 \tag{6.13}$$

This allows the reliability engineer to reduce the testing time, but care must be taken to ensure that no other failure mechanism is activated due to the accelerated stress levels. For example, if the fatigue life is N_3, the failure mechanism would change from high-cycle to low-cycle fatigue during the

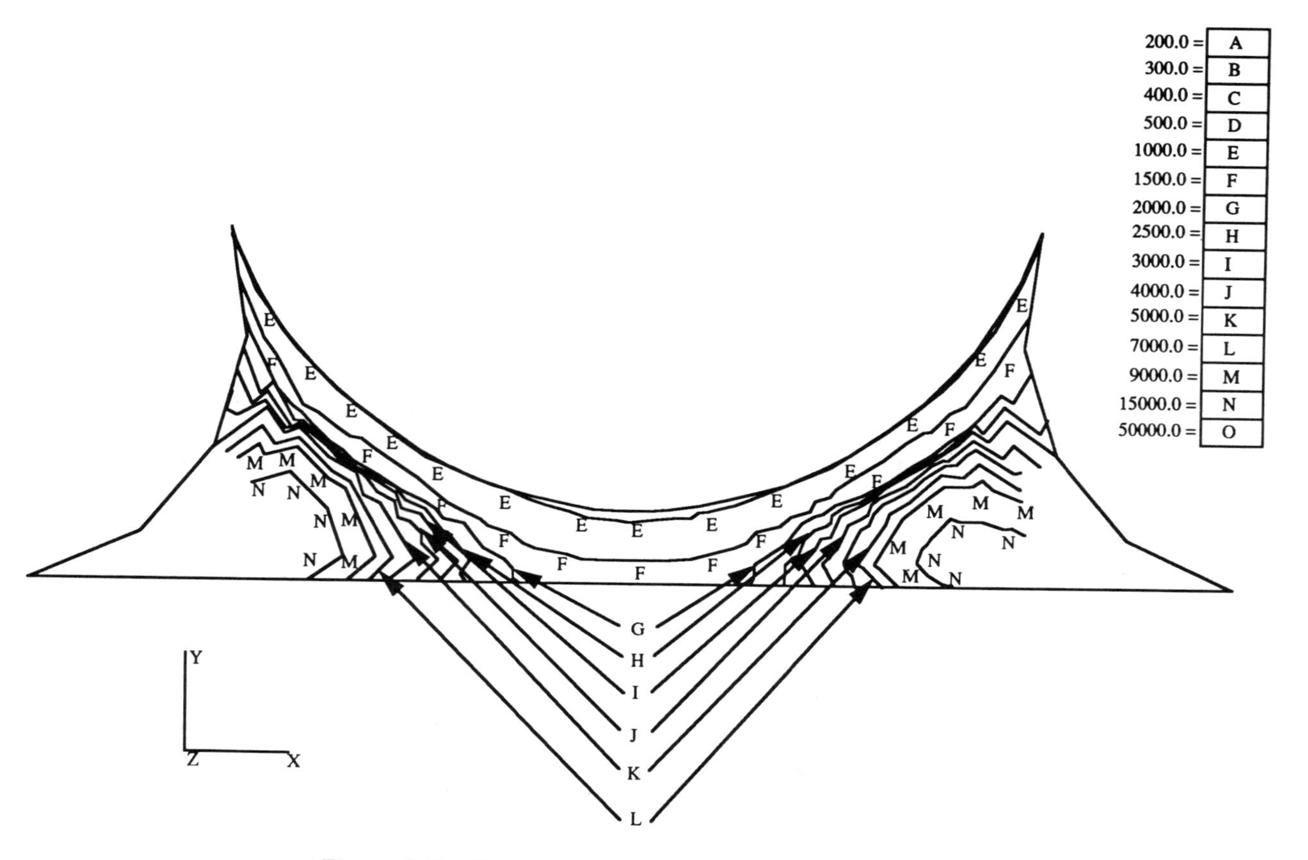

Figure 6.11: Contour plot of the life of the solder joint

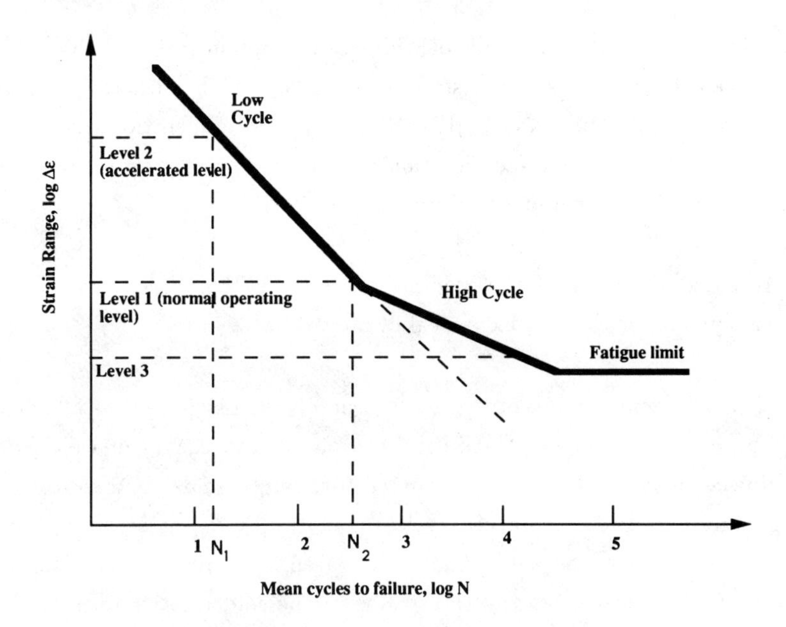

Figure 6.12: Accelerated testing for fatigue life

accelerated test. The extrapolation

$$N_3 = N_2 \times \text{Level}\,2/\text{Level}\,3 \tag{6.14}$$

is no longer valid. Realistic worst-case usage environment and accelerated testing conditions are well documented in Engelmaier [1991a,b].

The primary parameter to accelerate a solder-joint test is a decrease in the dwell time, though strain can also be increased unrealistically to force failures. Increasing temperature is not an effective option, because it activates a myriad of other failure mechanisms. However, test samples can be created with a deliberately large CTE mismatch, thus increasing the strain levels with a relatively lower change in temperature. Mechanical cycling can also be applied, but the correlation between isothermal strains and thermal strains due to temperature cycling is controversial. Solder generally experiences stress relaxation during fatigue cycling, while the dwell time is inevitably reduced during accelerated tests. It is, therefore, essential that the dwell time during testing be well calculated to allow for the required creep and stress relaxation [Engelmaier 1983, 1985]. The difference is fatigue damage, which is accounted for by an acceleration transform, details of which can be found in Engelmaier [1989]. A low-acceleration test has a transform of 10 to 20; a high-acceleration test has a transform between 100 and 500 [IPC-SM-785]. Low-acceleration tests provide a "benchmark" result; high-acceleration tests have a greater error margin and are generally used for comparitive analysis.

Accelerated testing of solder joints can be done by thermal cycling, mechanical cycling, or thermomechanical cycling.

Thermal cycling is performed by alternately stressing devices at hot and cold temperature extremes. The critical test parameters that regulate the damage induced in the solder joint are temperature range, solder temperature rate changes, number of temperature cycles, and dwell time at temperature extremes. The test can be used at an accelerated level by increasing the temperature range, thus increasing the strains in the solder. A temperature rise beyond 20°C/min. becomes a thermal shock test that does not allow time-dependent deformation. Severe temperature gradients can distort the printing wiring board; this distortion introduces a multiaxial loading on the solder joint, thus creating a different problem. Thermal shock is, therefore, not recommended as an accelerated test for solder joints. The solder experiences constant temperature loading in many applications, but the dwell time in the accelerated tests can be reduced. The result is reduced stress relaxation and, therefore, less fatigue damage during the test; this damage can be calculated to ascertain the acceleration factor of the solder joint.

Mechanical cycling is effected by applying cycles of displacement or cycles of load on a solder attachment. The test is conducted at a constant temperature. The variables include the load or displacement range applied, the loading or displacement rate, the test temperature, the method of load or displacement application (tension, shear, or bending), and the hold time employed.

Thermomechanical cycling employs temperature and mechanical loads simultaneously. The temperature is cycled while a mechanical load is applied: for example, cycling can be performed with tensile loading while heating, and with compressive loading while cooling. These tests are very complex and can be correlated to the field failures.

6.4 Quality assurance

Quality assurance of solder joints is an ongoing task associated with the monitoring of materials and process parameters to ensure that previously qualified parameters continue to be maintained within the specified tolerances. Quality assurance is not intended to check the ability of the solder joint to survive its design life, but to ensure that all variabilities beyond a specified tolerance (defects) are weeded out from the product population. Defects could be due to any one or a combination of the following factors,

- **Poor solderability, dewetting, and non-wetting.** These refer to poor surface characteristics of the printed circuit board pads, component leads, metallizations, impurities in the solder or flux, cleaning material, and aging of the surfaces.

- **Intermetallics.** These problems are caused by excessive heating during soldering, or aging of the solder joint at high temperature for a long time.

- **Process parameter deviations.** These involve issues such as deviation in equipment settings, temperature profiles, depth of immersion, conveyor speed and angle, maintenance, and other such factors [Manko 1986].

Quality involves understanding all the early failure processes and checking nominal process parameters with continuous on-line verification and control in every production lot. Verification essentially ensures that the attributes of each product are within validated tolerances. Section 6.5 covers details on the physics of early failures, while Section 6.6 discusses some existing approaches to assure quality, such as statistical process control (Section 6.6.2) and screening (Section 6.6.1). Efficient monitoring is needed, as part of the manufacturing process, to obtain reliable solder joints.

6.5 Early failure issues

Early failures result from improper assembly and manufacturing, which introduce defects and residual stresses into the solder joint. These flaws are "stress raisers" that act as potential failure sites where cracks can initiate and cause the ultimate failure of the device. Understanding the causes and physics of failure mechanisms and modes and the relevant load analysis helps to ensure high yield, quality and reliability; a clear understanding of the influence of process parameters on the introduction of defects helps the design team formulate robust manufacturing process guidelines.

6.5.1 Solderability, dewetting, and non-wetting

The integrity of a solder joint is qualified by the term "solderability." The term is not well defined in the literature, but can be explained as the ability to solder easily and effectively. Solderability depends on such factors as wettability, soldering temperature, flux, and impurities. Some causes of poor solderability include:

- surface deposits on the lead or substrate, which prevent melting or alloying of the coating;
- improper heat distribution in the solder;
- incorrect surface coating;
- impurities or co-deposited material in coating;
- dirt at the coating or the base metal or underplating interface;
- oxide at the coating or the base metal or underplate interface;
- improper cleaning or plating procedures;
- intermetallic formation during storage;
- and metallic overcoat without sufficient bond to base material.

The soldering industry has tried to simplify these issues, relating solderability to wetting forces and wetting time [Manko 1986] and vice versa [Wolverton 1991]. However, the true relationship between solderability and wettability is yet to be determined. While wettability can be quantitatively measured, good wettability is necessary but not sufficient to ensure good solderability.

Wettability affects solderability through non-wetting and dewetting. Non-wetting results when there is a physical barrier between the base metal and the solder. The cause of non-wetting is usually impurities — any material on the base metal which is resistant to the flux or soldering conditions used [DeVore 1989]. Often low temperature does not allow the solder to melt and wet properly and can cause non-wetting. The impurities result in either a poor intermetallic bond between the solder and the base metal or no bond, both of which decrease the contact area between the solder and the bond pad. The non-wet zones act as stress raisers and potential sites for fracture initiation.

Many of these problems can be solved by ensuring a high-purity work environment, carefully screening and selecting components, and using the correct flux. The defect introduced by non-wetting is latent and extremely hard to detect.

Dewetting is a stage between complete wetting and total lack of physical bonding caused by the cohesive forces in the solder material, which pull the solder back, creating minute areas of non-wetting. This phenomenon resembles the behavior of oil on a steel substrate, coalescing into small lumps and leaving a very thin layer wetting the metal. Dewetting may also involve the emission of gas during exposure of the part being soldered to molten solder. The source of the gas can be thermally degraded flux or hydration released by it [DeVore 1988, 1989]. The degree of dewetting depends on the amount of gas, its composition, and its location.

Wettability is critical for good solder joints. The extent to which the solder wets the base metal and the time taken for wetting are both important. Generally, the wetting time should be short, because a longer wetting time indicates a nonreducible film on the surface being soldered [Romig et al. 1990]. The wetting rate is therefore used as an indication of the wetting characteristics of the lead (see Section 6.6.1). Wetting, in turn, depends on the materials and flux used, the viscosity of the solder, the thermal reactions taking place, the heat transfer, and the lead geometry. Wetting force and visual criteria are other means for measuring wettability.

The wetting by a liquid of a solid surface is expressed in terms of the interfacial energies of the solid-liquid-vapor combination, as given by Young's equation [Thwaites 1982]:

$$\gamma_{LV}\cos\theta = \gamma_{SV} - \gamma_{SL} \tag{6.15}$$

where γ_{LV} is liquid-vapor interfacial energy, γ_{SV} is solid-vapor interfacial energy, γ_{SL} is solid-liquid interfacial energy, and θ is contact angle.

The amount of wetting and spreading of molten filler alloy depends on the free energies of the various surfaces involved in the system. Equation 6.15 represents a simple balance of forces acting on the interface between flux and basis metal, using the combined values of the filler or basis-metal interface. If the interfacial tension at the filler or basis-metal interface decreases due to the metallurgical alloying process, a greater force spreads the alloy over the surface, improving wetting. Ideal wetting results in a contact angle close to zero, which corresponds to infinite spreading. In practice, such a situation is difficult to achieve because other factors, such as surface roughness and chemical reaction, also have an influence. Young's equation assumes that no chemical reactions occur between the liquid and the solid substrate and that the solid substrate is absolutely smooth, these assumptions are seldom operable.

Wetting phenomena can also be modeled by the free-energy approach, which assumes that for one material to wet another, the resulting combination should have a lower total interfacial energy [Thwaites 1982]. Therefore, materials with lower surface energy (e.g., liquids) spread over materials with higher surface energy, and result in a lower net energy. Heinrich et al. [1990] have presented a two-dimensional solder-joint model, based on surface tension theory, which estimates the joint profile in terms of solder properties.

Degrees of wetting

Wetting is often categorized into degrees of wetting: fully wet, non-wet, partially wet, and dewet. The degree of wetting is associated with the value of the contact angle.

- **Fully wet.** The solder completely wets the base metal and the contact angle is approximately equal to zero.
- **Non-wet.** The contact angle has not formed and is greater than 90 degrees.
- **Partially wet.** The contact angle is between 0 and 90 degrees.
- **Dewet.** The solder "sits" on the substrate, and can be easily peeled off, leaving a smooth surface. Dewetting is usually caused by extensive growth of an intermetallic or occurs after the solder-substrate reaction has produced significant impurities.

Reducing wetting problems

It is easier to prevent than to detect and remove wetting problems. Tests devised to quantify wettability (see Section 6.6.1) can be used to check the incoming components. One test might not be conclusive and may need to be supplemented with others.

Poor component wettability can, to some extent, be overcome by the use of more active fluxes, but denser component packing causes difficulties in flux-residue cleaning. Often poor wettability is solved by excessive heating during the effort to solder, but excessive heating results in chunks of Cu_6Sn_5 (an intermetallic formed between copper substrate and solder) suspended in solder. Nickel also forms intermetallics when subjected to extended heat treatment; nickel-tin intermetallics can grow from a nickel barrier in the form of prisms, and can fracture and drift in solder [Parent et al. 1988]. Usually there is a storage period between the manufacture of the component and its assembly. Wettability can deteriorate during storage due to the formation of either an intermetallic between the substrate and the coating or oxide on the coating. The problem can be reduced by either applying a protective coating to the component or storing it in an inert atmosphere, which reduces surface degradation considerably but is expensive and difficult.

Coatings can dissolve in the solder, fuse, or do neither. A common, relatively inexpensive method is to use a fusible coating of tin or a tin-lead alloy. For this to work, the base metal must be carefully cleaned and checked for oxides or deposits before plating. Coatings such as gold, silver, cadmium, and zinc can also dissolve into the solder. The coating protects the substrate and dissolves in the molten solder when heated, exposing the clean surface to form a perfect bond. However, the dissolution can also have detrimental effects on the solder joint. For instance, if the gold-plated film is less than 1.3 microns, the film is porous and the solid does not act as an effective oxygen barrier. If the gold coating is greater than 3.2 microns, extensive intermetallics can form in the solder joint. The intermetallic is very brittle and can be detrimental to solder-joint reliability [Fox et al. 1986].

6.5.2 Intermetallics

Intermetallics are a necessary and natural outcome of the interaction that occurs in the soldering process, and some intermetallics are necessary to ensure a good bond between the solder and the substrate. During soldering, molten solder comes in contact with the substrate. As a result, the base metal dissolves into the individual constituents of the molten solder and the active constituents in the solder combine chemically with the base metal, forming intermetallic compounds on the surface of the base metal [Romig et al. 1990].

Inadequate intermetallics may indicate a poorly wetted or weak solder joint. Conversely, excessive brittle intermetallics may indicate inadequate ductility in the joint, leading to early fatigue failure. The relative amount of the base metal that dissolves into the solder can be estimated by the solubility curves of the base metal in the molten solder; the amount of intermetallic formed depends on the chemical affinity of the active metal in the base metal. Solubility and chemical activity are unavoidable when metals come in contact at high temperature; in the case of soldering, where interfacial surface tensions are altered in favor of improved wetting, solubility and chemical activity are useful.

A wide variety of intermetallics are formed depending on the solder and the substrate composition. These compounds generally tend to be more brittle and have a different co-efficient of thermal expansion than do the metals or alloys from which they are formed. Fields et al. [1991] presented the critical properties of bulk Cu_6Sn_5, Cu_3Sn and Ni_3Sn_4, which are summarized in Table 6.1.

The research conducted on quantifying the effect of intermetallics on the reliability of solder joints has been inconclusive. The general observation has been that the toughness of the solder joint decreases with an increase in the amount of intermetallics. Parent et al. [1988] evaluate the thickness of intermetallics and the debonding strength over time. They noted that a decrease in strength is accompanied by an increase in the amount of the intermetallic from 230° to 300°C, but there is very wide variation in debonding strength. Furthermore, the correlation of debonding strength with fatigue strength is also not clear.

Temperature cycling experiments have revealed that cracks sometimes develop prematurely and propagate through the intermetallic and the solder immediately adjacent to it [Frear et al. 1989]. The weaker of the two regions, solder or intermetallic, causes failures. The crack takes no definitive path while propagating to failure. (Figure 6.13 summarizes the potential path of the crack at the interface.) Furthermore, examination of failed solder joints has shown that brittleness is caused by the formation of shrinkage voids, which divert the fracture path away from the intermetallic-to-solder interface through the solder itself [Tomlinson and Bryan 1986].

Intermetallic compounds can be formed either by liquid-solid reactions or by solid-state diffusion. Intermetallic growth can occur at the interface between

Property	Cu_6Sn_5	Cu_3Sn	Ni_3Sn_4
Vickers Hardness (kg/mm^2)	378	343	365
Toughness (MPa $m^{1/2}$)	1.4	1.7	1.2
Young's Modulus (GPa)	85.56	108.3	133.3
Poisson's Ratio	0.309	0.299	.33
Thermal Expansivity (ppm/C)	16.3	19.0	13.7
Thermal Diffusivity (cm^2/sec)	0.145	.240	.083
Heat Capacity (J/gm/deg)	0.286	.326	.272
Resistivity (micro ohm-cm)	17.5	8.93	28.5
Density (gm/cc)	8.28	8.9	8.65
Thermal Conductivity (watt/cm-deg)	0.341	.704	.196

Table 6.1: Critical properties of Cu_6Sn_5, Cu_3Sn and Ni_3Sn_4

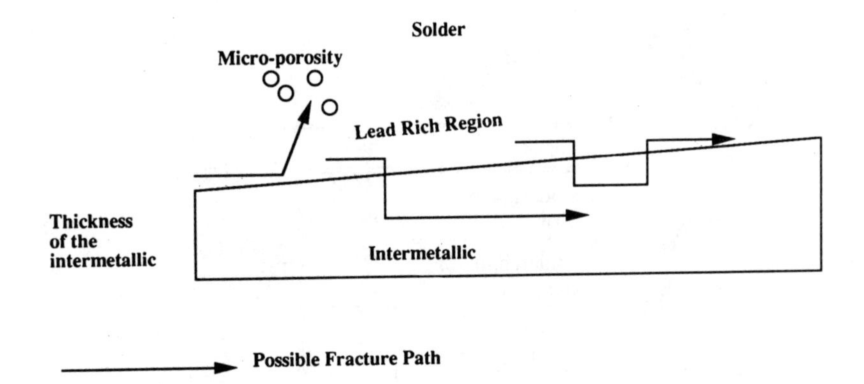

Figure 6.13: Possible fracture path in a solder joint

the substrate and the substrate coating after lengthy storage. The growth kinetics depend on the temperature and length of time the material is stored. The amount and type of the intermetallics govern the solderability of the assemblies. Impurities, such as sulfide, carbonates, and oxides, complicate the structure of the intermetallic and degrade the quality of the layer. Mechanical operations exposing the solder joint to oil, acid, and solvent vapors form additional complex compounds at the intermetallic and weaken the solder joint.

6.5.3 Other causes of early failures

Solderability and wettability defects account for more than 70% of defective solder joints. Other defects are caused by deviations of process parameters, poor maintenance, aging of equipment, negligence, or operator oversight. A typical ranking of defective solder joints is shown in Table 6.2.

Often problems can be solved by good management, efficient organization, and streamlined communication. However, the importance of analyzing manufacturing processes to reduce and control inherent variation is critical. The ability to recognize the symptoms of a defective solder joint and correctly diagnose the precise fault in the system comes with extensive experience. Troubleshooting guidelines can be found in soldering books such as those by Lambert [1988] and Manko [1986].

The most common defects and their causes are summarized in Table 6.3,

Defects	Ratio
Solderability of boards	50
Solderability of components	20
Handling and assembly	15
Design defects	5
Solder machine	5
Others	5

Table 6.2: Ratio of causes of defects in solder joints [Woodgate 1988]

but there are numerous other causes, not listed in the table, that are dependent on the manufacturer. Some of these defects include inconsistencies in printing fine-pitch solder paste, inconsistencies in the specification of the components, and process variation, such as reflow profile and handling.
Many of the resulting defects might not be defined in standard specifications. These require effective screening and process control strategies to ensure detection and prevention.

6.6 Measures to assure quality

Myriad tools can assure quality; these are detailed in any standard text on quality assurance [Raheja 1991]. The key measures to assure quality covered in this section are screening and statistical process control.

6.6.1 Screening

One important quality assurance measure is screening. The purpose of screening is to eliminate premature failures in the final product and increase yield. The process of screening involves early detection or precipitation of latent defects through appropriate methods. Some solder joints exhibit a decreasing rate of failure during the early period of their service life, typically resulting from the use of defective materials, poorly controlled manufacture and assembly processes, or mishandling. These joints contribute to the infant mortality (early failure) of electronic devices. Screening helps to reduce or eliminate early failures in field by detecting solder-joint defects, either by non-stress screens, like visual inspection and other non-destructive evaluation methods, or by stress screens, which involve the application of stress (not necessarily service stresses) to precipitate defects, often on an accelerated basis.

Stress screening

The intent of stress screening is to discover or precipitate defects and flaws that might manifest themselves as premature failures if they remain undetected. The screen does not simulate temperature cycling or vibrational loads, but stimulates early failures in solder joints; this is the most important difference between screening and testing. Stress screens are designed to cause minimal or no damage to good solder joints, and do not precipitate failures that would not

DEFECT	CAUSE OF THE DEFECT
Poor wetting and non-wetting	Non-wetting is commonly caused by contaminants on the board or the solder material, oxide or other chemical impurity on the surface, and other surface imperfections and material/environmental impurities.
Dewetting	Outgassing by the metal substrate, gas resulting from breakdown of flux used during soldering and cohesive forces in the solder
Disturbed (cold) solder joint	Vibrations cause displacement during the freezing stages of joint formation; caused by conveyor belt, bearings, pumps, ventilation or exhaust fan, etc.
Incomplete fillets, and poor solder rise	Poor solder rise can be caused by solderability problems in the solder pad or the component lead, misregistration of solder resist on the pad, uneven wave or conveyor vibrations, and poor fixturing. Sometimes in J-leaded devices or PLCCs, the solder paste climbs up on the component leads due to the component leads reaching the soldering temperatures first, leaving the leads void of solder
Excess solder	This can occur due to variation of the process parameters during soldering - e.g., incorrect depth of wave immersion, poor solderability of the board or the component, insufficient oil in the intermix, contaminated solder, and incorrect wave angle or speed.
Icicling	This can occur due to lack required temperature during the soldering process, due to heat sinks thermally connected to solder joints.
Bridging	This is an effect of excess solder and is caused by poor drainage of solder during soldering, conductors being bent or placed too close to each other, impurities in the material, or low heat balance.
Solder and component short circuits	Solder shorts occur due to misalignment, heavy paste deposits, or misaligned paste depositions.

Table 6.3: Common defects and their causes

DEFECT	CAUSE OF THE DEFECT
Electromigration, and tin whiskers shorts	This process is accelerated with foreign contaminants and moisture.
Current leakage	This is caused by the presence of ionic and non-ionic contaminants with humidity and other conductive materials.
Grainy or dull solder joints	These joints can result from dross mixed in the solder, intermetallic compounds, impurities in solder, variation in freezing patterns or high purity tin-lead, and insufficient heat.
Formation of solder balls	Detached balls of solder form during reflow when the paste is oxidized, preventing complete flow of the solder onto available wettable surfaces of the substrate and the chip carrier. Extremely difficult to remove, these balls can cause shorting problems if dislodged. Causes of solder ball formation could be: insufficient oil in the flux; dried up flux; uncured solder mask; failure of the solder to peel off the PWB surface; moisture in the solder paste; contamination on the PWB surfaces; fast temperature ramp up in reflow operation causing, exploding the solder paste; improper baking of the solder paste prior to reflow; inactivaity of flux system due to excessive shelf life.
Voids in the solder joint	Voids can be introduced into the solder joints by the flux, oxides, and other non-metallic material in the solder pastes. They are introduced by inadequate reflow time, boiling point of the solvents, paste composition, flux activation, and conductor pad metallization.
Component tilting (tomb stoning)	The component tilts up from the pad when the adhesive is sparsely applied, or applied over a dirty surface; or in reflow soldering uneven melting causes one joint to wet before the next and surface tension tilts the component.
Webbing	Webbing can be caused by improper curing of the resin in the laminate or the solder resist, a rough surface resulting from excessive scouring, or flux starvation.

Table 6.3: Common defects and their causes (contd.)

occur under normal field use. Screening introduces minimum wear-out damage [Environmental Stress Screening of Electronic Hardware 1990].

To be effective, stress screens must be designed and tailored for a specific failure mechanism; for example, the screen to weed out poor wetting defects has different trigger parameters than the screen to eliminate excess intermetallic defects. The defects are processing-technology-specific.

The initial step in the stress screening process is to identify the potential flaws affecting the solder joints and the responsible failure mechanisms. The screen(s) is then selected to accelerate these defect mechanisms. Most of the defects in solder joints cause early fatigue failure due to stress raisers, brittle solder joints, and other defects (discussed in Section 6.5). Temperature and mechanical cycling are among the most common stress screen techniques.

The most critical step of the stress screening process is to determine the screen and screening parameters — dwell time, rate of temperature increase, and temperature range of the temperature cycling test. These parameters are specific to the defect (voids, dewetting) and to the solder joint (gull-wing, J-lead). An incorrect value of the parameter can damage the solder joint, fail to precipitate the defect, and render the screens totally wasted. There is no one way to select screen parameters; a combination of the following techniques can be used to select and specify the screening parameters.

- **Finite element modeling.** The effectiveness of the parameters can be investigated by using finite element modeling techniques, which involve incorporating the defect while modeling the solder joint and studying its effect on the joint. Figure 6.14 shows the decrease in life, as calculated by finite element analysis, of a misaligned J-leaded surface-mount solder joint [Dasgupta et al. 1992b]. Different screening parameters can be simulated using similar techniques, and the optimum point can be determined.

  Modeling is an effective option when the physics of the defect are well understood. Unfortunately, all defects cannot be modeled; in some cases, the physics of the defect are not yet understood (for example, it has not yet been ascertained how a crack grows in the interface between the solder and the intermetallic), and often the material properties are unknown.

- **Error seeding.** Another technique is "error seeding", or subjecting a known number of solder joints with defects leading to infant mortality

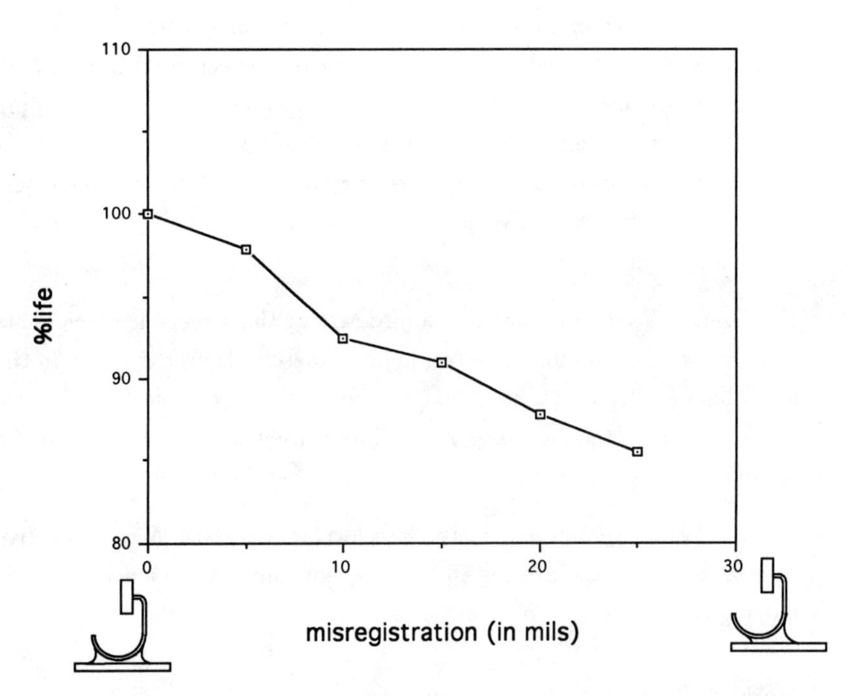

Figure 6.14: Effect of misregistration on life of a J-leaded surface-mount solder joint life

to the screening process. The failure mechanisms and stresses that stimulate the defects are also known. Stress magnitudes are determined by progressively increasing the screen regimens until a known percentage of defects in the test sample of solder joints have been precipitated. Stress levels are thus fixed to screen out these known defects. The defects in the test sample must be measurable and controllable for reliable experimental results. For example, dewetting can be studied by varying the area of the wettable copper pad and by covering the pad with a layer of unwettable metal, like copper oxide. The proportion of non-wettable oxide to the wettable pad indicates the degree of wettability. Screens can be used to study the dewetting defect on these solder joints.

- **Step-stress analysis.** Step-stress analysis is a common technique to establish stress levels for screening defective solder joints. In the analysis, progressively stronger stresses are imposed on the solder joint, which is subsequently checked for defects. As each failure occurs, failure analysis is conducted to determine the cause; if it is due to a latent defect and not to overstress failure of a good solder joint, the stress level is further increased. The process continues until a defect is precipitated by overstress failure which does not occur in field conditions. The upper limit of the stress level is thus determined for the defect precipitated. A large enough test sample is required to ensure that the screen can filter out all of the possible associated latent defects.

- **Feedback.** Feedback from the failure rate of the screened solder joints can also be used to monitor screening parameters. Defect tracking in the assemblies and careful failure analysis are required. The effect of changes in screen parameters on the rate of failure is used to optimally adjust the screens.

- **Cost.** Cost is a dominant factor in selecting a screen. A cost-effective screening program addresses all the relevant failure mechanisms by employing a minimum set of screens.

Non-stress screening

The "cosmetic" appearance of the solder joint is not a sufficient criterion for assuring the reliability of a surface-mount solder joint [Leas 1991]. In fact, appearance can be very deceptive. For example, resoldering an indium solder joint with tin-lead solder can result in a clean appearance that masks intermetallics of inferior fatigue resistance [Marshall 1991]. This section discusses the some of the methods for non-destructive evaluation (NDE). NDE is potentially the most effective tool for assessing solder joints, because no solder is consumed during testing. As the industry moves towards fine-pitch technology, destructive evaluation has presented practical problems due to its size. NDE methods, which evaluate more than just surface cosmetics, include two- and three-dimensional reflectance, and thermal and x-ray imaging techniques [Millard 1989].

The reflectance system is like an optical microscope, and uses lenses to visually aid the inspector. The magnification can vary, but the system is subjective. The equipment can be made more sophisticated by using a camera to produce an image on a monitor. The image is stored in pixels and is compared with the available database for automated detection of defective solder joints; the computer vision reduces subjectivity. Some equipment uses lasers, which reflect light from the assembly to examine the solder joints. The light is received by sensors and processed by computer to obtain a three-dimensional representation of the image. Three-dimensional imaging provides the advantage of quantitative data about the geometry and location of solder volume with respect to the pad.

Another class of NDE methods uses Nd-YAG laser to input thermal energy to examine the integrity of solder joints. Radiating infrared energy heats the solder joint and an infrared detector focused on the solder volume monitors the cooling profile. The output is stored as a computer image that is compared with the image database stored in the memory. Thermal systems can be used to rapidly capture images.

X-rays can also be used to scan a substrate to evaluate solder joints. The X-rays are generated by a high-voltage source, which travels through the substrate and bombards a fluorescent screen. A digitized image is produced by a video camera, and image-processing techniques are used to identify solder-joint defects. In transmissive radiography, one such method, the source and the detecting screen are fixed and the substrate is positioned in the desired field of view. The transmissibility of the X-rays depends on the material content of the viewing field. For example, excess intermetallics in the solder produce an image very different from that of a solder joint with a higher percentage of solder.

Adjustments can be made in the setup to focus on the joint at different viewing depths, allowing the interior of the joint to be examined. However, double-sided solder joints are difficult to examine using the X-ray technique, as the data become too complex to analyze.

Laminographic radiography, a recently emerging technology, uses an X-ray source and a fluorescent screen in synchronized motion that allows a focus on only one area at a time, thereby overcoming problems with examining double-sided boards. Stress or non-stress screening is an expensive process and adds no real value to solder joints. The manufacturer should use screening as a tool to provide feedback to improve the manufacturing process.

6.6.2 Statistical process control

When manufacturing solder joints in bulk, a 100% screening program may be neither cost effective nor necessary. When screening mature products, it may be advisable to follow a statistical sampling technique, in accordance with existing statistical process control (SPC) procedures. The sample size and frequency are based on the actual defect rate, reliability models and risk analysis.

For outgoing solder joint quality it is very important to produce joints that conform to target specifications. Variability of the manufacturing process, materials, and assembly cause variability in solder joints, and thus poor quality. SPC can be used to ensure that the parts are being built to a specific process target throughout the manufacturing cycle. The direct result of SPC manufacturing is product predictability through process control. If process and product targets are centered and distributed well within a product specification, direct benefits include fewer rejects, smooth manufacturing flow, improved yield, and improved prediction of yield [Motorola 1990]. Furthermore, inspection of all solder joints on the board to ensure that only four defective joints are produced in a million (six-sigma requirement) is very expensive. The high cost of solder-joint inspection, combined with the increasing difficulty of fine-pitch lead repair (see Chapter 7), requires implementating of the SPC system. However, with the advent of automatic inspection tools, there is the potential to apply SPC in the manufacture of defect-free solder joints [Mani et al. 1992].

Statistical process control employs tools for studying variation in a manufacturing process and for monitoring its stability. It involves inspecting or measuring certain quality parameters or attributes on a sample lot of the manufactured products. The potential quality parameters include the following [Swenson 1991]:

- left, right, and toe fillet mean solder height;
- left, right, and toe fillet mean pad-wetting angle;
- left, right, and toe fillet mean fillet-top wetting angle;
- left, right, and toe fillet mean fillet curvature;
- left, right, and toe fillet mean fillet solder volume;
- left, right, and toe fillet distortion pad wetting angle, distortion (variation of pad/fillet wetting angle measurements); and
- left, right, and toe fillet distortion top wetting angle, distortion (variation of fillet/lead wetting angle measurements).

The magnitude of the parameters can be plotted in control charts, which can be used to detect if the process is out of control and plan steps to bring the process back into the specification limits. Control charts indicate when the process is going out of control. There are two basic types of control charts: control charts for variables (X-bar charts, range charts) and control charts for attributes (p charts, c charts). The generic steps in statistical process control:

- evaluation of process behavior by means of control charts;
- determination of process capability;
- a function of process variation and the tolerance of the quality parameter under consideration;
- control or elimination of the variables that affect the process; and
- monitoring of control charts and corrective actions before defective parts are produced.

Solderability testing is a common process control check performed while surface-mounting components. Sample coupons may be taken out of an assembly line mounting process to check for solderability.

Existing methods by which wetting parameters of the leads can be tested include the dip test and the wetting balance test. In the dip test, the specimen is immersed in a temperature-controlled solder bath. The rate of immersion, dwell time, and withdrawal are carefully controlled, according to the ranges recommended in IPC-S-805A. Mechanical devices are preferred for dipping the specimen to prevent wobble, vibration, and other undesirable movements. The solder is allowed to cool in air; the specimen is then examined for 95% or more solder coating over the surface.

The wetting balance test involves dipping the specimen in a temperature-controlled molten solder bath and recording the dipping force as a function of time. The dwell time, immersion, emersion rate, and flux used are all specified. The recorded pattern of wetting force must cross the corrected zero axis at or before one second of test time at a standard ramp rate and dwell. Leadless components require a special holding fixture to verify the solderability of the solderable ends.

6.7 References

Attarwalla, A.I., Tien, J.K., Masada, G., and Dody, G. Confirmation of Creep and Fatigue Damage in Pb/Sn Solder Joints. ASME Winter Annual Meeting

(1991).

Barker, D., Dasgupta, A., and Pecht, M. PWB Solder Joint Life Calculations under Thermal and Vibrational Loading. Unpublished.

Coffin, L.F., Jr. Fatigue at High Temperature, ASTM STP 520. Philadelphia, PA: American Society for Testing and Materials (1973).

Dasgupta, A., Oyan, C., Barker, D., and Pecht, M. Solder Creep-Fatigue Analysis by an Energy-Partitioning Approach, ASME Winter Annual Meeting (1991).

Dasgupta, A., Verma, S., and Barker, D. Fatigue Life of Misregistered J-Lead Solder Joints through an Energy-Partitioning Analysis, ASME Winter Annual Meeting (1992).

DeVore, John A. Solderability and Surface Mounting. *Circuit World* 14 (1988) 4.

DeVore, John A. Failure Mechanisms in Soldering. *Electronic Materials Handbook* 1, ASM International (1989).

Engelmaier, W. *Brazing and Soldering* 9 (Autumn 1985), 40.

Engelmaier, W. *IEEE Components Hybrids Manufacturing Technology Transactions*, CHMT-6 (1983).

Engelmaier, W. Environmental Stress Screening - Its Impact on the Reliability of SMT Solder Joint and Plated Through Holes. NEPCON West, *Proceedings of the Technical Program* 1 (1991a).

Engelmaier, W. Solder Attachment Reliability, Accelerated Testing, and Result Evaluation, *Solder Joint Reliability*, ed. J. Lau, New York: Van Nostrand Reinhold (1991b).

Environmental Stress Screening of Electronic Hardware (ESSEH). Environmental stress screening survey for assemblies, Proceedings of the 6th National Conference and Workshop, Institute of Environmental Science (November 1990).

Fields, R.J., Low, S.R., and Lucey, G.K. Physical and Mechanical Properties of Intermetallic Compounds Commonly Found in Solder Joints. Gaithersburg, MD: National Institute of Standards and Technology (1991).

Fox, A. et al. The Effect of Gold-Tin Intermetallic Compound on the Low Cycle Fatigue Behavior of Copper Alloy C72700 and C17200 Wires. *IEEE Transactions on Component, Hybrids, and Manufacturing Technology* CHMT-9 (September 1986).

Frear, D., Grivas, D., and Morris, J.W. Parameters Affecting the Thermal Fatigue Behavior of 60Sn-40Pb Solder Joints. *Journal of Electronic Material* 18 (1989) 6.

Goddard Space Flight Center. Evaluation Report, Solder Joint Fatigue Analysis. Serial No. 03347.

Halford, G.R., Hirchberg, M.H., and Manson, S.S. Temperature Effect on the Strain-Range Partitioning Approach for Creep Fatigue Analysis, ASTM STP 520. Philadelphia, PA: American Society for Testing and Materials (1979).

Hall, H.M. Creep and Stress Relaxation in Solder Joints, *Solder Joint Reliability*, ed. J. Lau., New York: Van Nostrand Reinhold (1991).

Heinrich, S.M. et al. Solder Joint Formation in Surface Mount Technology. *Journal of Electronic Packaging, Transactions of the ASME* 112 (September 1990), 3.

IPC-S-805A. Solderability Tests for Component Leads and Terminations (January 1989).

IPC-SM-785. Guidelines for Accelerated Reliability Testing of Surface Mount Solder Attachments. 6th Working Draft (December 1990).

Kapur, K.C., and Lamberson, L.R. *Reliability in Engineering Design.* New York: John Wiley and Sons (1977).

Lambert, Leo P. *Soldering for Electronic Assemblies.* New York: Marcel Dekker, Inc. (1988).

Lau, John H., and Rice, Donald W. Solder Joint Fatigue in Surface Mount Technology: State of the Art. *Solid State Technology* (October 1985).

Leas, Colin. Evidence That Visual Inspection Criteria for Solder Joints Are No Indication of Reliability. *Proceedings of NEPCON* (1991) 23-38.

Mani, B.S., Von Voss, William D., and Tong, Steven H. A Process Control Plan for Fine-Pitch SMT Assemblies. *Circuits Assembly* (January 1992).

Marshall, James L. Scanning Electron Microscopy and Energy Dispersive X-ray (SEM/EDX) Characterization of Solder - Solderability and Reliability in *Solder Joint Reliability*, ed. J. Lau., New York: Van Nostrand Reinhold (1991), 180-183.

Millard. *Electronic Materials Handbook* 1, ASM International (1989).

MIL-STD-883C. Test Methods and Procedures for Microelectronics. Washington, D.C.: US Department of Defense (1983).

Morrow, J. Cyclic Plastic Strain Energy and Fatigue of Metals. Proceedings of Symposium in Internal Friction, Damping, and Cyclic Plasticity, ASTM, STP-378 (1965) 45-87.

Motorola. Reliability and Quality Handbook. Motorola Semiconductor Products Sector (1990).

Oyan, Chen. Analysis of Thermomechanical Fatigue in Solder Joints. University of Maryland, College Park, MD: Masters Thesis (1991).

Parent, J.O.G., Chung, D.D.L., and Bernstein, I.M. Effect of intermetallic formation at the interface between copper and lead-tin solder. *Journal of Material Science* 23 (1988), 2564-2372.

Pecht, M., and Wen-Chang, Kan. Critique of Mil-Handbook-217E Reliability Prediction Methods. *IEEE Transactions on Reliability* 37 (December 1988), 5.

Raheja, Dev G. *Assurance Technologies - Principles and Practices*, New York: McGraw-Hill (1991).

Romig, A.D., Jr., Chang, Y.A., Stephens, J.J., Frear, D.R., Marcotte, V., and Lea, C. Physical Metallurgy of Solder-Substrate Reactions, *Solder Mechanics, A State of the Art Assessment*, ed. D.R. Frear, W.B. Jones, K.R. Kinsman, Warrendale, Pennsylvania: TMS Publications (1990).

Sandor, B.I. Life Prediction of Solder Joints: Engineering Mechanics Methods, *Solder Mechanics, A State of the Art Assessment*, ed. D.R. Frear, W.B. Jones, and K.R. Kinsman, Warrendale, Pennsylvania: TMS Publications (1990).

Shames, Irving H. and Cozzarelli, Francis A. *Elastic and Inelastic Stress Analysis.* Englewood Cliffs, NJ: Prentice Hall (1992).

Swenson, Raymond. Validation of an Automatic Solder Joint Quality Measurement System. *Surface Mount Technology* (October 1991) 25-34.

Steinberg, D.S. *Vibration Analysis for Electronic Equipment.* New York: John Wiley and Sons (1988).

Tien, John K., Hendrix, Bryan C., and Attarwala, Abbas I. The Interaction of Creep and Fatigue in Lead-Tin Solders, *Solder Joint Reliability*, ed. J. Lau, New York: Van Nostrand Reinhold (1991).

Thwaites, C.J. *Capillary Joining - Brazing and Soft-Soldering.* Research Studies Press, New York: John Wiley and Sons (1982).

Tomlinson, W.J. and Bryan, N.J. The Strength of Brass/Sn-Pb-Sb Solder Joints Containing 0 to 10% Sb. *Journal of Material Science* 21 (1986) 103-109.

Tribula, D., and Morris, J.W., Jr. Creep in Shear of Experimental Solder Joints. ASME Winter Annual Meeting (1989).

Woodgate R.W. *The Handbook of Machine Soldering.* New York: John Wiley & Sons (1988).

Wolverton, M. Component Solderability. *Circuits Assembly* (March 1991).

Chapter 7

Rework, Repair, and Manual Assembly

Louis A. Abbagnaro
PACE, Inc.
Laurel, MD

7.1 The repair process

While a perfect electronics environment where repair actions are never required has been a goal for many years, this level of perfection has yet to be achieved on an industry-wide basis. Many electronic modules go through rework or repair action at some time during their life cycles for a variety of reasons.

- Reasons for production repair include:
  - installation of the wrong component,
  - improper installation of the correct component,
  - defective components,
  - damaged substrates, and
  - design changes.
- Reasons for post-production repair include:
  - field updates,
  - component failures, and

– environmentally produced failures (thermal, electrical, etc.).

In addition, many component installation techniques developed for repair are used for manual assembly processes, such as engineering prototypes and short-run production.

Because rework, repair, and manual assembly are a part of the present electronics environment, it is important to understand these processes in order to select the proper equipment and the correct techniques to ensure success. Only with such an understanding can high-quality results be achieved rapidly and at minimum cost. Techniques that meet these criteria for both through-hole and surface-mount assemblies are described in this chapter.

7.1.1 Through-hole assembly and repair

Table 7.1 shows the general process steps for through-hole and surface-mount assembly and repair. For through-hole assembly, components are inserted into holes in a printed circuit board (PCB). These PCB holes stabilize the component until it is soldered in place. In a normal production process, wave soldering is used to mechanically fasten and electrically contact the components; manual soldering is typically used for short-run production. After soldering, the assembly is cleaned to remove flux and other remaining contaminants.

In a through-hole repair process, the last three steps in Table 7.1 constitute a "re-manufacturing" that mirrors the assembly process. The first three steps (desoldering the component leads, lifting out the component, and cleaning the assembly) constitute a "de-manufacturing" of the PC assembly. The combination of de-manufacturing and re-manufacturing defines the total repair process. Historically, through-hole repair was an unreliable process until the advent of the continuous vacuum desoldering handpiece, which made through-hole repair actions safe, productive, and cost effective.

7.1.2 Surface-mount repair

Table 7.1 shows the steps for surface-mount assembly and repair. Because the PCB no longer has holes, the components must be mounted and stabilized on the surface of the board until a reflow action can solder all connections. Stabilization is commonly achieved by applying solder cream (solder paste) to all contact areas of the PCB and then placing the components into the cream. The high viscosity of the cream stabilizes the components until reflow

COMPONENT MOUNTING	ORIGINAL ASSEMBLY	REPAIR
THRU-HOLE	- Insert component in hole	- Desoldering old component - Remove old component - Clean PC assembly - Insert new component
	- Wave solder	- Solder new component
	- Clean PC assembly	- Clean PC assembly
SURFACE MOUNT	- Apply solder paste	- Unsolder old component - Remove old component - Clean PC assembly - Tin pads & leads or apply solder paste
	- Target, align & place component	- Target, align & place new component
	- Reflow PC assembly	- Reflow new component
	- Clean PC assembly	- Clean PC assembly

Table 7.1: Through-hole and surface-mount assembly and repair process

occurs. Normally, reflow is achieved by either a vapor-phase, infrared, or forced-convection process. Alternately, a wave soldering process may be used; in this instance, however, the components are often bonded to the PCB with an adhesive to prevent misalignment during transit through the solder wave. Cleaning is typically the final step of the assembly process.

For surface-mount repair, desoldering is no longer suitable. Rather, the component is removed by heating all leads to solder melt temperature at the same time and then lifting the component off the PCB. If the component was attached with adhesive during production, the removal process must include a shearing action to break the adhesive bond. Surface land preparation, during which the excess solder is removed, is the next step. The specific techniques used for reflow, removal (unsoldering), and surface-mount land preparation vary with the component being removed. As with the through-hole process, once the surface-mount assembly has been de-manufactured, it must be re-manufactured using steps that again parallel new production. Although production equipment may be used for the re-manufacturing process, it is more frequently accomplished with manual or semi-automated repair equipment and techniques.

7.1.3 Manual assembly and repair environments

Normally, a production environment is highly automated to efficiently manufacture large quantities of high-quality printed circuit assemblies. The repair environment, by contrast, is typified by change. The required repair action may vary from one assembly to the next, and more than one repair may be needed on a single assembly. The repair may occur at any point in the life of a PCB, from initial manufacturing to a field modification several years later. Equipment for repair tasks is normally designed for versatility or flexibility, rather than for dedicated repetitive action, to meet the variety of problems that may be encountered and to accommodate the different areas where the repair action is likely to occur.

7.2 Assembly and repair goals

7.2.1 Process considerations

Assembly and repair are carried out using three processes:

- fully automated,
- semi-automated, and
- manual processes.

The ideal goal for all these processes is the same: excellence. In modern terminology, excellence can be defined in many ways — for example, zero defects or six sigma. In practice, something short of excellence is often achieved the wrong repair process is selected or the correct process is inadequately implemented. The better the choice of a correct process and its implementation, the better the chance of achieving excellence.

The major difference in the three types of assembly processes is the degree of human involvement. There is little human action in a fully automated process, but there is significant involvement in manual and semi-automated processes. These processes must therefore be designed to achieve excellence using techniques involving people. Often, the addition of the human factor into the assembly or repair process dramatically influences the choice of techniques that may be used safely.

7.2.2 Classifying repair processes

Repair processes may be classified as indicated below.

- **Component-level repair.** This process removes damaged or incorrectly installed components and replaces them with correctly installed parts. It is the most frequently encountered repair action and has become significantly more complex with the advent of surface-mount technology.

- **Circuit-level repair.** In this process, a portion of the printed circuit assembly is replaced or refurbished, usually to restore electrical continuity. This is often performed in conjunction with component-level repair.

- **Modification.** Modification changes printed circuit assembly to add a new electrical function or to correctly implement an originally included function. Modification includes tasks such as adding jumpers or new components.

- **Salvage.** Salvaging is done when good components must be removed from an otherwise unusable assembly.

The goal of all repair processes should be excellence. Virtually no repair action is highly automated today. Rather, manual or semi-automated equipment is most often used. This requires a high degree of human involvement to determine the correct action to take and then to select the proper equipment and techniques to execute it successfully.

7.2.3 Achieving excellence in repair

A satisfactory repair requires four elements:

- establishing a controlled process;
- satisfying mechanical and thermal safety criteria;
- using properly trained operators; and
- achieving reasonable productivity rates.

Establishing a controlled process

Process control is well understood in manufacturing environments, but is often ignored in conjunction with repair. Simply stated, process control occurs during a repair action if there is a defined set of steps that trained operators with reasonable skill can perform to yield repeatedly acceptable results. Because a major portion of most repair actions involve the use of manual tools, process control is only achieved if the correct tools and techniques are used. The tools must be correct in terms of safety (see Section 7.2.3), as well as from a user standpoint. The best-conceived process does not work if the operator is unable to carry it out properly. Some repair techniques may only be safe if the operator is aided in some critical steps by a semi-automated technique using a machine — for example, to place a component or control the focusing and delivery rate of a hot-gas or infrared heat source.

Thermal safety considerations

While there is some flexibility in manual assembly and repair actions and often more than one technique can be used, some fundamental safety criteria must be satisfied for success.

Through-hole guidelines.: Although manual assembly and repair of through-hole components have been performed for years, many changes have

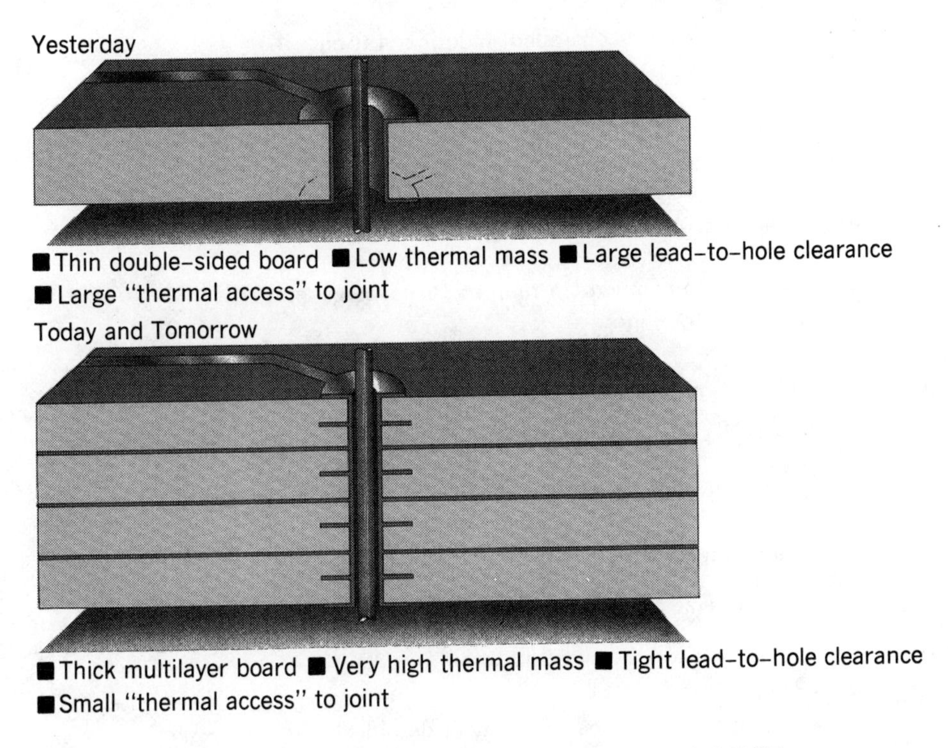

Figure 7.1: Typical cross-section of a double-sided PCB

occurred, as indicated by the cross-section views of the two through-hole PCBs shown in Figure 7.1. The printed circuit assembly has changed from a single-sided or double-sided topology to a multilayered design. The multilayer circuit is typified by smaller pad areas and greater thermal mass within the assembly because of internal power and ground planes. It thus takes more heat delivered through a smaller heat conduit to achieve solder melt on a multilayer PCB when soldering or desoldering.

While more heat is required, the area where the heat is delivered is usually smaller and fabricated from thinner copper laminate material. A common action on older PCBs was to raise the temperature to handle heavy loads or to simply work faster, but an indiscriminate application of heat causes lifting of the lands, or burning or measling of the substrate. With modern multilayer assemblies, temperatures above 400°C (750°F) are usually lethal to PCBs. Useful through-hole thermal safety guidelines for selecting equipment for through-hole manual assembly or repair include the following:

- Use a temperature-controlled soldering station.
- Use the tip that best fits component lead size.
- Use the lowest tip temperature that gives good productivity.
- Use the appropriate soldering times — typically 2 to 5 sec. per joint.
- Employ recommended tip temperatures for general soldering — 320° to 400°C (600° to 750°F).
- rNote that recommended tip temperatures for desoldering are 340° to 400°C (650° to 750°F).
- Consider alternatives to temperatures > 400°C for heavier loads:
  - auxiliary heating on component side;
  - larger size or different shape tip with greater surface area;
  - addition of solder or flux to aid heat flow.

A temperature-controlled soldering or desoldering station is a critical component. These so-called "closed loop" heating systems are designed to sense and respond to load requirements, even while operating at relatively low-idle tip temperatures. The only way an open-loop system can deliver heat at maximum capacity is to operate at maximum tip temperature, which is often 535°C (1000°F) or more. These high initial tip temperatures can no longer be tolerated by modern PCBs.

The safe delivery of heat is the single most critical factor to consider in through-hole assembly or repair actions. One of the other important considerations related to safe heat transfer is the selection of the proper tip. In the past, it was desirable to select one tip that could be used in most, if not all, soldering applications. Today, soldering or desoldering tips should be selected with specific guidelines in mind.

- The tip with the largest surface area that will fit the work being done should be selected, and the tip should be changed when work changes. Heat transfer to the workpiece is usually limited most severely by tip geometry. For example, a relatively fine-point conical tip should not be used for general-purpose soldering. The small tip surface area limits heat transfer and usually prompts the operator to turn up the temperature unduly to overcome the problem.

- Once the tip is selected, work should be done at the lowest practical tip temperature. Usually, temperature is governed by the loss of productivity. Human operators have a natural speed in repetitive operations of between 2 and 5 sec. If the tip-and- temperature combination is working too slowly, an operator typically increases the soldering iron temperature. Interestingly, the normal human reaction time for soldering and desoldering is about the optimum time to form ideal solder joints with hand tools. At too high a temperature, a human reacts just as fast, but more heat than is necessary is transferred to the solder joint, increasing the chance of damaging the assembly. Figure 7.2 shows solder extraction of a joint at two different tip temperatures. The operator worked just as fast, whether the tip temperature was 400°C 750°F) or 345°C (650°F); however, appreciably more heat than was necessary was transferred at the higher temperature. Instead of arbitrarily raising tip temperatures, operators should attempt to achieve a balance of high productivity at the lowest possible temperature.

The best through-hole soldering and desoldering stations can meet the challenge of most modern PCB assemblies. The selection process for new equipment should ensure high-quality temperature control and should compare the ability of the equipment to operate effectively at temperatures below 400°C (750°F).

Surface-mount safety guidelines. Surface-mount assembly and repair also require the safe application of heat; major considerations for the removal or installation of a surface-mount component are summarized in Figure 7.3, which shows a typical surface-mount assembly. The objectives are to (1) remove the main component while not overheating it, (2) avoid overheating the substrate area (to prevent blistering or measling), and (3) avoid overheating adjacent solder joints or adjacent components (refer to Figure 7.3.

These primary objectives can be further elaborated into the specific considerations indicated below.

- Maximum component internal die temperature:
  - normal components: 220°C (430°F)
  - sensitive components: 150°C (300°F)

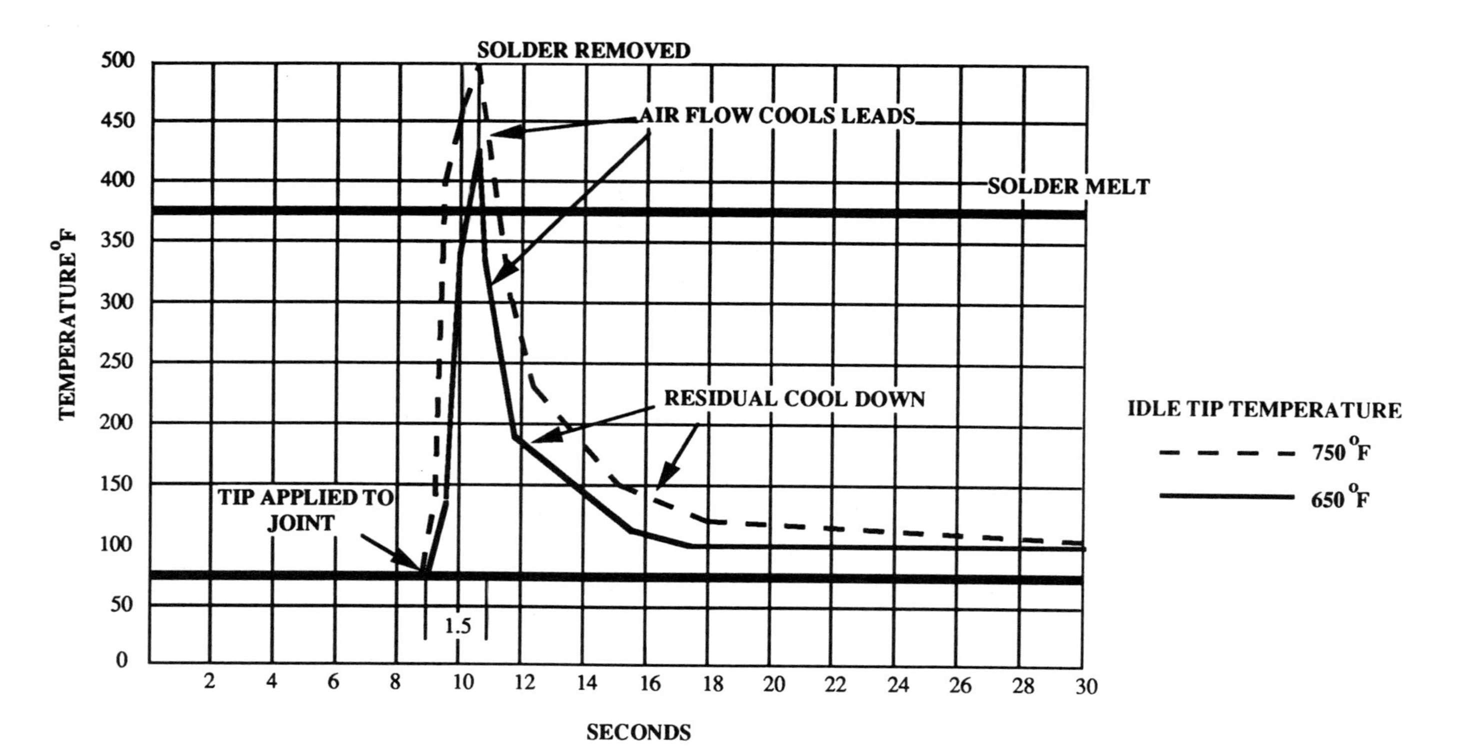

Figure 7.2: Thermal profile of a plated through-hole during a desoldering process at different tip temperatures

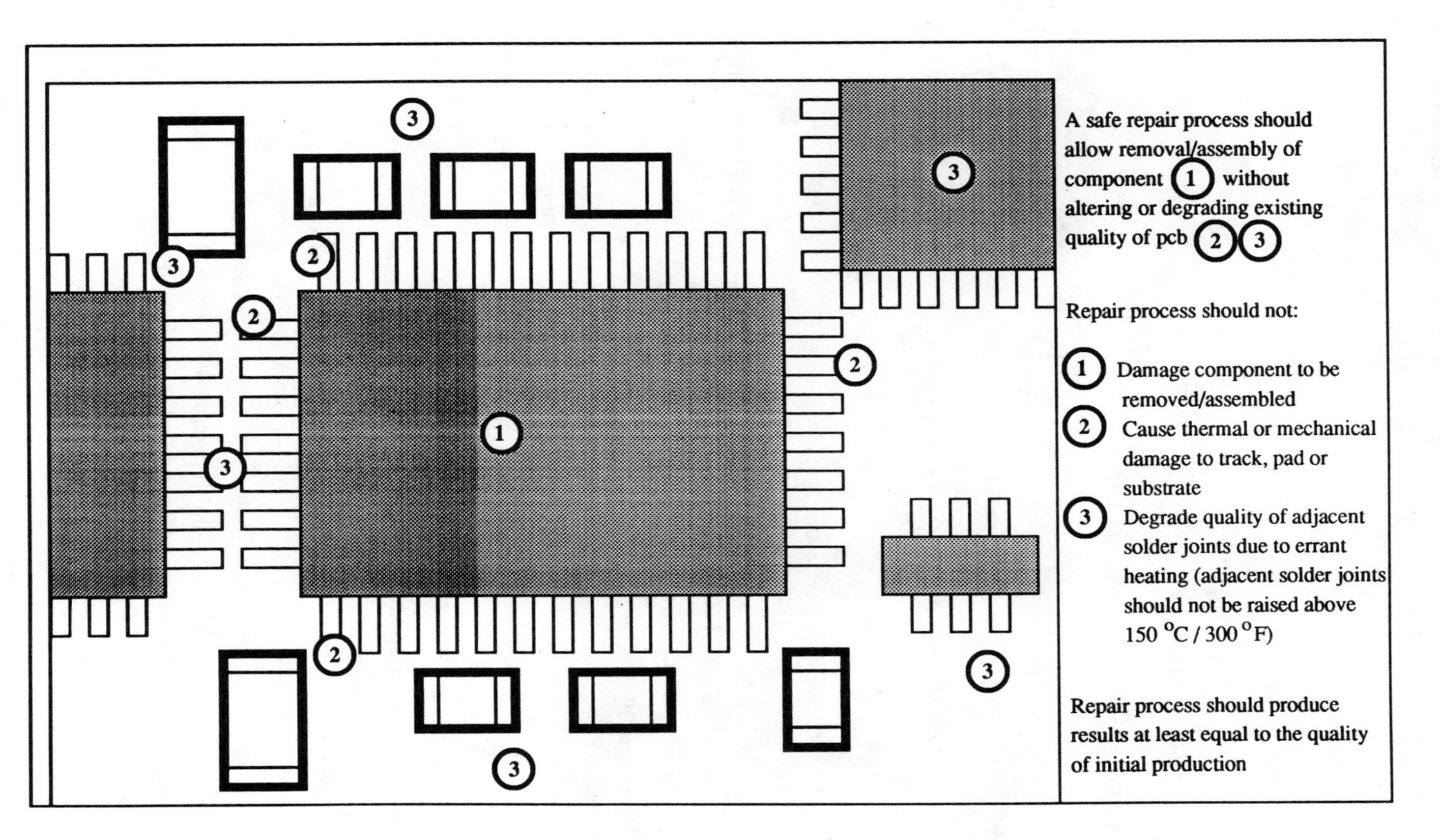

Figure 7.3: Layout of a surface-mount PCB showing major manual assembly and repair concerns

- Solder joint formation:
  - temperature: 220°C (430°F)
  - time: 2 to 5 sec. final reflow
  - cooling rate: approximately 100°C (210°F in 1 min.)
- Adjacent component solder joints:
  - maximum solder joint temperature: 150°C (300°F)
- Thermally sensitive components (ceramic capacitors):
  - thermal ramp rates: typically 2° to 6°C (4° to 11°F) per second
- Ceramic substrates:
  - preheating temperature: 100° to 125°C (210 to 250°F)

The internal temperature for components is based on temperatures equivalent to those occurring during primary production reflow processes. If a thermally sensitive component is used that cannot handle temperatures as high as 220°C, special production processing is needed. For example, the part might have been assembled using a thermode (hot bar) method to keep die temperatures below 150°C. Solder-joint formation temperatures of about 220°C with dwell times of about 3 to 5 sec. yield maximum-strength solder joints in most situations.

One unique aspect of manual assembly or repair is the need to avoid overheating adjacent components or their solder joints. Some components, such as ceramic chip capacitors, are sensitive to the rate of heat-up; furthermore, the ramp rates must conform to manufacturers' specifications to avoid damage. Excessive reheating of existing solder joints weakens the integrity of these joints and can cause premature failure. The need to concentrate heat delivery is not new to manual assembly and repair; in the past it has been guaranteed in through-hole assemblies by using conductive tools that transfer heat only to areas where the heated tip touched the workpiece. The same guidelines have led to the development of many effective tools and techniques for surface-mount assembly and repair.

Depending on the heating method selected, other general guidelines for safe surface-mount assembly and repair, listed below, may be applicable.

- For conductive (contact) heating:
  - Only temperature-controlled heating methods should be used. The reason is the same as for through-hole devices: namely, the ability to work at the lowest possible operating tip temperatures and still have sufficient heat-transfer capacity to rapidly reflow all the leads simultaneously.
  - Whenever possible, tools and tips should be selected that deliver the heat directly to the lead or land (solder joint) areas. Proper tools and tips usually give the safest and quickest means to remove a component.
  - Solder and flux should be used as they are in through-hole applications to aid the transfer of heat. They should be used in tinning tips, fluxing highly oxidized joints, and even prefilling leads or using solder preforms to facilitate heat transfer and removal.
  - Whenever possible, hand tools that permit one-hand component removals should be used. The lifting technique should be incorporated into the handpiece design. In this way, components can be removed as soon as reflow has occurred; otherwise a component must be heated well past reflow to allow time for inserting a tool to lift the component from the PCB before the solder solidifies.
- For convective (gas) heating:
  - Focused heat delivery is critical. The main danger with convective heating is the lack of directional control (focusing) of the gas flow. This limitation can lead to damage of adjacent solder joints during a reflow action, even with trained operators.
  - Control of the heat delivery rate is critical. The governing factors are gas temperature and gas flow rates. It is usually easier to overheat and burn a substrate with hot gas than with conductive tools because the rate of heat delivery is often not load-responsive.
  - Hot gas should be supplied only when needed. Some simple systems have no means of starting and stopping the gas flow. The continuous flow of hot gas makes assembly or repair tasks more difficult to complete safely.

These factors suggest the use of simple hand-held hot-gas tools primarily for installations. With simple single-point or bifurcated nozzles, an operator can focus the tool properly. For component removals, where all sides must be heated simultaneously, semi-automated hot-gas machine systems (Section 7.5.4) are recommended. These systems are designed to guarantee the proper delivery of gas at the right temperature to the desired areas of the PCB for the correct amount of time.

- For infrared (IR) heating:
  - IR assembly and repair equipment is available generally only in machine versions. The main guideline is to select systems that focus the light at the lead or land areas. Early IR systems illuminated a large image around the component to be installed or removed and saturated the area with heat. This saturation technique creates two serious problems: adjacent components are overheated during the process, and the IR energy heats the component faster than the solder joint because of differences in light absorption characteristics. The component and underlying substrate often get so hot that serious warping or damage occur. Most modern IR stations now provide the equivalent of a hot-gas nozzle to focus the light energy, minimizing the potential for damage with an IR approach.

Very recently some specialized systems using intense halogen or xenon lamps or lasers have been used for specific surface-mount assembly actions. These techniques offer promise for future component repairs.

Using properly trained operators

With manual and semi-automatic assembly or repair procedures, true process control and safety can be achieved only if the operator both has the proper equipment, and also understands which technique to use, what tools to select, and how to use these tools. Too often, the major cause of poor quality in manual assembly and repair actions is the lack of operator training. In better-managed companies, a strong emphasis is placed on training and monitoring operators performing assembly and repair tasks. Operators should be taught proper techniques and given suitable practice time before attempting repair of good assemblies. They should also understand a few fundamentals of heat transfer to avoid inadvertently causing thermal damage.

Achieving reasonable productivity

Controlling the process and safety are necessary, but many production or service managers cite the obvious axiom, "time is money". Many assemblies are scrapped rather than repaired if repair time is too long or repair cost too high. Thus, the best assembly and repair techniques must provide reasonable speed as well as safety. The definition of "reasonable" depends on the cost of the specific assembly. For a very expensive PCB, an intricate repair process taking several hours might be worthwhile. With most PCBs, only relatively quick repair processes of less than a few minutes are acceptable. In later sections, representative times for manual installation and removal processes are indicated. These are useful guides for deciding about whether to repair or scrap a specific assembly. When equipment is selected, testing the time for repair actions should be an obvious step; however, it is frequently overlooked until repair equipment has been purchased.

7.3 Additional considerations

7.3.1 Single- and double-sided through-hole assemblies

Generally, these are the easiest through-hole assemblies for manual installation and removal of components. Hand soldering and desoldering have been used for many years. The safest techniques should use the lowest temperatures consistent with good work productivity. Modern soldering and desoldering stations should be able to operate safely on most single- or double-sided assemblies at tip temperatures below 345°C (650°F).

7.3.2 Multilayer through-hole assemblies and components

As indicated in Figure 7.3, manual assembly or repair of multilayer circuit boards poses new challenges. Again, good general guidelines are to use the lowest temperatures possible and to avoid soldering and desoldering tip temperatures above 400°C (750°F). Assembly and repair are often more difficult in a multilayer assembly because of heavy ground planes. The worst situation occurs when a solid ground plane is connected to a lead on the component side of the board, often inhibiting the complete melting and flow of solder through a plated through-hole in assembly operations and inhibiting the complete removal of solder in repair operations. The only safe way to overcome this problem is to use auxiliary heat on the component side of the circuit board; the common solution of using temperatures well above 400°C produces lifted lands or burned substrates. Perhaps the most difficult through-hole component to install or remove when there are heavy ground planes is a pin-grid array (PGA). In the PGA, leads tied to power or ground planes are often not accessible, making auxiliary heating difficult to achieve, if not impossible. Auxiliary heat can sometimes be supplied through the component case if it is made of a ceramic-based material.

7.3.3 Surface-mount components and assemblies

The size or shape of most through-hole components is relatively unimportant to the assembly and repair task; only the number and size of leads to be soldered or desoldered are considerations. Because surface-mount components require simultaneous reflow for all removals and some installations, component case and lead design are very important. The physical differences in surface-mount components have prompted the development of many new techniques to install and remove these devices; a soldering iron or solder extractor alone is no longer sufficient. At least two additional distinctive conductive handpieces have been developed to aid surface-mount repair actions. For removing PLCCs and LCCs, a thermal tweezer handpiece best meets the thermal delivery criteria mentioned in Section 7.2.3. For flat packs, a thermal pick handpiece that contains a heater with a built-in vacuum pick also meets the criteria. A soldering iron equipped with surface-mount tips can safely remove chip and SOIC components. While the function of all these devices is to heat all leads simultaneously and then lift the part, differences in lead and case configuration of the parts they are designed to remove dictate distinct designs.

7.3.4 Thermally sensitive components and assemblies

Most surface-mount components and assemblies can withstand the sudden application of heat to their lead structures during the removal and installation process. In some situations, such as with ceramic body components and other thermally sensitive component types, the heat must be delivered slowly to prevent damage to the component or PCB. When assembling or repairing these components, it is usually necessary to preheat the component or assembly slowly to a temperature about 50°C below final reflow. The rate of heat ramp-up is typically between 2° and 6°C/sec. Manufacturers' specifications should be checked carefully before installing or removing any thermally sensitive parts.

Ceramic substrates can also be damaged if heat is applied too quickly, so often these assemblies are preheated to minimize thermal shock. Preheat is also required with ceramic PCBs, which dissipate heat rapidly, making reflow difficult to achieve with primary heat alone. As a general guideline, preheat temperatures of 100°C to 125°C are normally recommended for ceramic substrate preheating. The heat-up should be gradual over a 5- to 10-min. period. Temperature-controlled hot plates, such as the one shown in Figure 7.4, are ideal for this purpose.

Generally, techniques described in the following sections may be used with ceramic components and PCBs, if preheating is also employed.

Figure 7.4: Hot plate preheating a ceramic PCB

7.4 Through-hole component repair techniques

In the following sections, techniques that normally are considered thermally safe are described and typical thermal profiles are shown.

7.4.1 Desoldering

The primary method used to remove through-hole components is continuous vacuum desoldering, as shown in Figure 7.5. Although simpler extraction techniques used in conjunction with a soldering iron are still used, they do not provide all the elements needed for good desoldering. A satisfactory desoldering process should be able to do the following:

- heat the solder rapidly, using as low a tip temperature as possible;
- move the lead to ensure that the solder has melted;

Figure 7.5: Through-hole desoldering

- evacuate the molten solder rapidly, using a quick-rise vacuum while continuing to move the lead;
- flow cool air through the aspirated hole for 1 to 2 sec. to prevent a resweat joint while continuing to move the lead.

The thermal guidelines discussed earlier suggest the use of typical temperatures of 345°C (650°F) for desoldering and maximum temperatures of 400°C (750°F). Rapid vacuum is the other factor necessary to extract solder cleanly. The absolute level of vacuum is not particularly important, but the increase in vacuum from ambient to about 250 mm of Hg (10 in.) in 200 ms or less is very important to good desoldering. This ensures that all the solder is removed in one slug. A slow vacuum build-up generally leaves solder in the hole to reattach the lead.

The best clue to solder melt is the ability to move or "wiggle" the lead. Often, the trained operator eliminates this step from the process. The direction of the wiggle depends on lead geometry (Figure 7.6a) — back and forth for leads with a rectangular cross section, and orbitally for leads with a circular cross section (Figure 7.6b). Without moving the lead, the operator cannot know whether the lead is ready for solder evacuation. If the lead does not wiggle in 3 to 5 sec., the tip should be removed from the joint and the operator should consider using either a higher temperature or auxiliary heat. If the tip temperature must be raised above 400°C (750°F), auxiliary heating of the lead from the component side of the PCB is recommended to avoid thermal damage.

Finally, cool air should be drawn through the evacuated hole for about 2 sec. to lower the temperature of the hole and lead below solder melt, thereby preventing the formation of a resweat joint. The thermal action of the desoldering process shown in Figure 7.2 indicates the rapid cooling of a joint when air flows through the evacuated hole during a proper extraction.

Simple solder pullers do not provide this air flow and require excessive heating to ensure that the solder is molten before extraction. This technique is still in use, but is not considered suitable for safe extraction. Units with built-in vacuum and air flow should be used for best results in through-hole extraction. Typically two versions are available today: (1) extraction stations with built-in vacuum pumps, and (2) stations that use shop air to develop vacuum. Either type works properly if it provides proper heat control, fast vacuum rise time, and sufficient air flow. These key performance areas should be

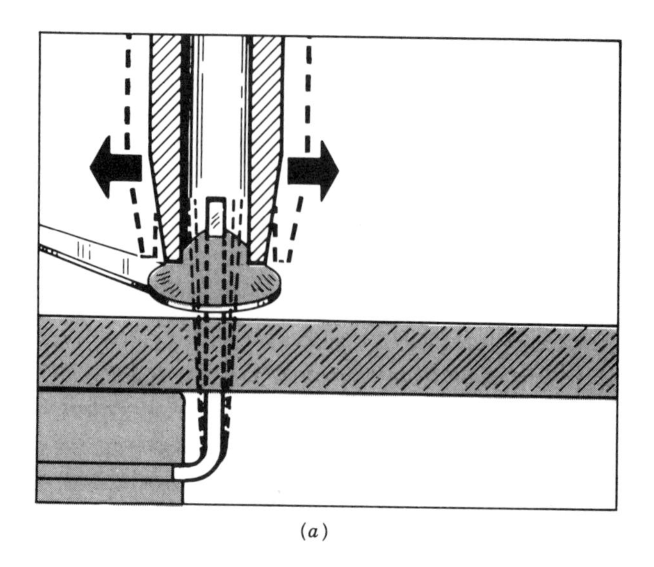

(a)

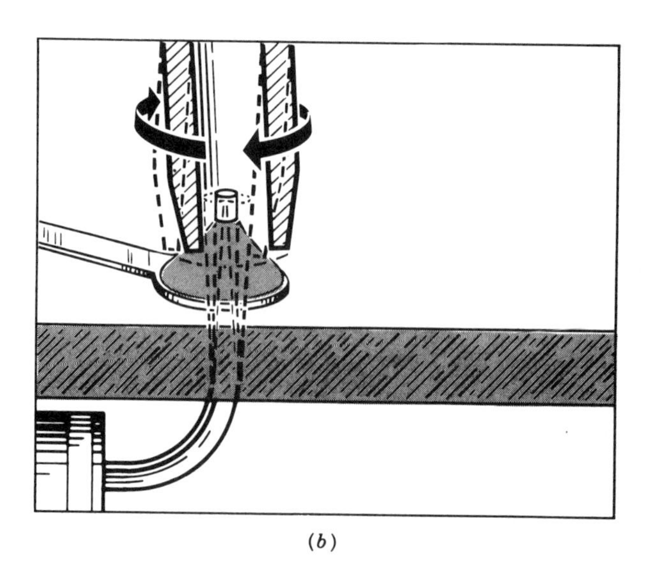

(b)

Figure 7.6: Wiggling the leads in desoldering

compared prior to any equipment purchase. Because the wiggling and targeting of the extractor tip over a lead are also critical elements, extractor handpieces should be compared to determine which designs permit the easiest targeting and manipulation. Good manipulative ability makes the operator's task easier and improves the quality of the desoldering.

Through-hole soldering

Thermal guidelines for modern through-hole soldering are given in Section 7.2.3. A critical decision is selecting a station that permits safe soldering of modern assemblies. Soldering stations may be categorized as follows:

- open-loop (constant power) stations with fixed or variable power;
- closed-loop (temperature-controlled) stations with:
  - fixed temperature,
  - preset selectable temperature,
  - variable-temperature dial display,
  - variable-temperature electronic display, and
  - microprocessor-controlled electronic display.

Open-loop soldering stations. The oldest form of soldering iron equipment, many of these stations are still used today. The fixed-power irons are usually plugged into an AC outlet and heated to a specified temperature. The lack of temperature control may not be a problem for non-critical repairs, such as cable or connector assemblies, but is a severe limitation for work with modern through-hole or surface-mount PC assemblies. Fixed-power open-loop soldering stations should never be used on modern PCBs.

Variable-power controllers permit changing the tip temperature of the open-loop soldering iron by increasing or decreasing the power delivered to the heater. Controllers used for this purpose are no more complex than simple light dimmers. The most serious disadvantage of this type of controller is that full power can only be delivered to the soldering iron when the unit is operating at maximum temperature (often $> 500°C$). At lower temperatures the power delivery to the iron is dramatically reduced. As a result, the recovery of the iron is slowed, limiting its ability to handle repetitive tasks uniformly, especially under heavy thermal loads.

Closed-loop soldering stations. Closed loop or temperature-controlled stations use some form of thermal sensor to detect heat at or near the tip. The sensor signal is then used as part of a feedback circuit that adjusts heater power to maintain the temperature at a desired set point. The major advantage of closed-loop stations is that the initial tip temperature can be set to match the sensitivity of the load, thereby preventing damage. Once heat transfer to a load starts, the heater power increases up to its maximum capacity to match the thermal requirements of the load. Thus, the closed-loop station is better suited to safely handle modern PCBs, and is the only station that should be considered for electronic assembly or repair.

The simplest closed-loop system, the fixed-temperature design, operates at a single temperature (usually between 370° to 425°C), useful for general work but often either too high or too low for many specific tasks. The major advantage of this design is that an operator cannot arbitrarily increase temperature and cause thermal damage to an assembly.

A step up from the fixed-temperature design is a station that can be operated at two or three preset temperatures. Often temperature presetting is achieved by changing either a plug or a tip heater cartridge to select the desired temperature. Again, the major advantage of this design is that it limits the opportunity for inadvertent operator error and thermal damage. With the variety of tip and load requirements encountered today, preset temperature soldering stations are often restrictive.

The most versatile closed-loop stations have a control knob to permit temperature adjustments over a relatively wide range. In most stations, a calibrated temperature control dial is provided, although a few variable control stations are uncalibrated. More expensive stations provide an electronic readout which usually indicates temperatures with a 1° resolution.

Finally, the most sophisticated stations available today use microprocessor technology to provide temperature control. These stations often permit an accurate knowledge of the true tip temperature with any style of through-hole or surface-mount tip. They provide other desirable features, such as temperature lock-out and automatic shut-off when the unit is not in use for a predetermined period.

Electrical and thermal safety are also considerations when selecting soldering stations or, for that matter, selecting any other repair equipment. The accepted norm for this equipment in the U.S. today is MIL-STD-2000, which requires soldering equipment to meet the following safety specifications:

- tip resistance to ground: < 5 ohms;
- tip leakage: < 2 mV RMS;
- idle tip temperature stability: ±5°C (±10°F);
- zero voltage switching circuitry to avoid transients; and
- static safe handpieces.

Most high-quality soldering stations available today meet MIL-STD-2000, but users should verify EOS/ESD suitability prior to use.

Other through-hole component extraction methods

A few techniques are available for the removal of such through-hole components as DIP packages. Typically, they use a thermal tweezer handpiece to heat all the leads and then pull out the package. The apparent advantage of this approach is speed. For example, at 3 sec. a joint, a 20-pin DIP requires 1 min. to desolder. A thermal tweezer extraction of the same part may take only 10 sec.

The thermal tweezer extraction, however, may not always produce the best results. With a thermal tweezer, the solder remains in the through-holes making it difficult to insert a new component without either removing the remaining solder (desoldering) or using the same thermal tweezer to reheat the residual solder when a new part is installed. If solder extraction is used, it should be employed from the start. If a new part is inserted into holes containing old solder, the resulting joint is usually weaker and more susceptible to failure than when fresh solder is used.

Therefore, the complete removal of through-hole components using thermal tweezer techniques is often not recommended. It is perfectly acceptable, however, if the operator is only trying to salvage components from an otherwise unusable assembly.

7.4.2 Semi-automated techniques

A few specialized pieces of equipment are available today for automated installation and removal of through-hole components.

Solder fountains

This equipment is specifically designed to remove and install large through-hole components, such as connectors. Typically, the component is placed over a small solder wave or fountain that heats all leads of the component simultaneously. The part is then lifted from the PCB. Often, pressurized air is quickly blown through the holes to remove much of the residual solder.

During installation, the fountain flows fresh molten solder into all the holes and the part is manually inserted. The board is then removed from the fountain with the part resoldered. The molten solder fountain generally ensures that fresh solder is added to all the holes and overcomes most of the objections to the manual thermal tweezer process described in Section 7.4.1.

Solder fountains often work well and save a great deal of time when large parts are installed or removed. However, problems are sometimes encountered because heavy ground planes in boards make solder reflow difficult. These same ground planes often cause the circuit board to warp when the PCB is heated to a high temperature. Operators should evaluate solder fountain equipment to ensure that it provides a safe and controlled process.

Robotic soldering

Many repetitive production soldering operations may be automated using robotic soldering stations. These systems are often used to install parts that cannot travel through a normal wave-soldering operation. Typically, a robot arm manipulates a conventional soldering iron with a solder feeder attached. The arm is programmed to handle a specific soldering task, such as attaching all the leads on a large connector. The use of robotic equipment, versus conventional hand soldering, is typically dictated by cost or quality considerations.

7.5 Surface-mount component repair techniques

This section describes techniques currently in use for the removal of surface-mount components that meet the safety criteria discussed in Section 7.2.3. Thermal profiles are provided for several of the applications.

7.5.1 Manual removal techniques

To ensure thermal safety, all of the following techniques should use temperature-controlled handpieces, as recommended for soldering irons. Surface-mount tips require an additional warning: because of the large physical size of many surface-mount tips, a considerable differential in temperature exists between the "set" temperature on the dial or digital display of the soldering station and the "actual" temperature of the tips. Figure 7.7 show why this difference occurs. In most temperature-controlled stations, the sensor is near but not exactly at the heated tip, producing a difference in the temperature of the tip and that at the sensor. This difference, or "offset," is usually negligible with through-hole tips (Figure 7.7) because of their small size. With larger surface-mount tips, tip offsets are significant and cannot be ignored. The tip offset depends on the size of the tip and the operating temperature. Figure 7.8 shows tip offsets for a few surface-mount tips; tip offsets can exceed 50°C on larger tips. Most soldering stations do not provide compensation for surface-mount tip offset and indicate temperatures significantly higher than actual tip temperatures. A few manufacturers provide for tip offsets so that their stations can indicate the true tip temperature. When this information is not provided, an individual calibration is necessary to determine the true tip temperature.

PLCC/LCC removals

PLCCs and LCCs are readily removed using a thermal tweezer, such as shown in Figure 7.9. The thermal tweezer tips can be designed to conform to the lead geometry, ensuring heat delivery at the solder joints and providing fast removals at relatively low tip temperatures. Thermal tweezers often use tip temperatures below 300°C (575°F). (Table 7.2 shows typical removal times for many different types of components, including PLCC removals using a thermal tweezer.) The other advantage of the thermal tweezer is that a single device both lifts and heats, permitting one-hand action and greatly simplifying the removal task. Thermally, the process is quite safe, as indicated in Figure 7.10, which shows the critical temperatures at the component solder joints, internal die, and adjacent solder joints during the removal of a 68-pin PLCC.

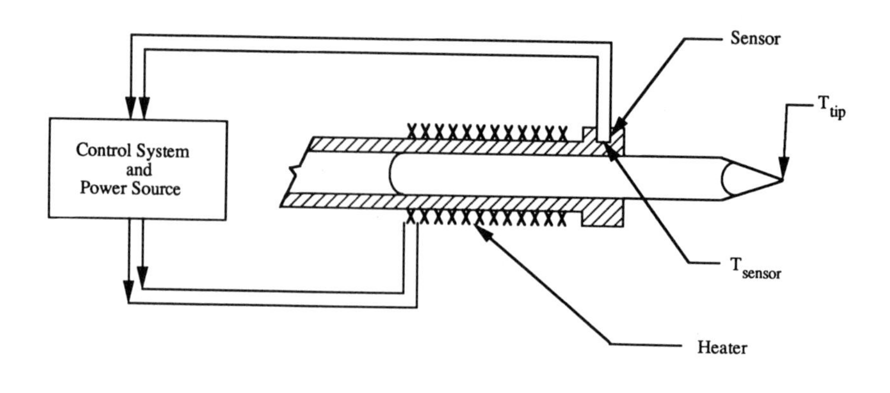

(a)

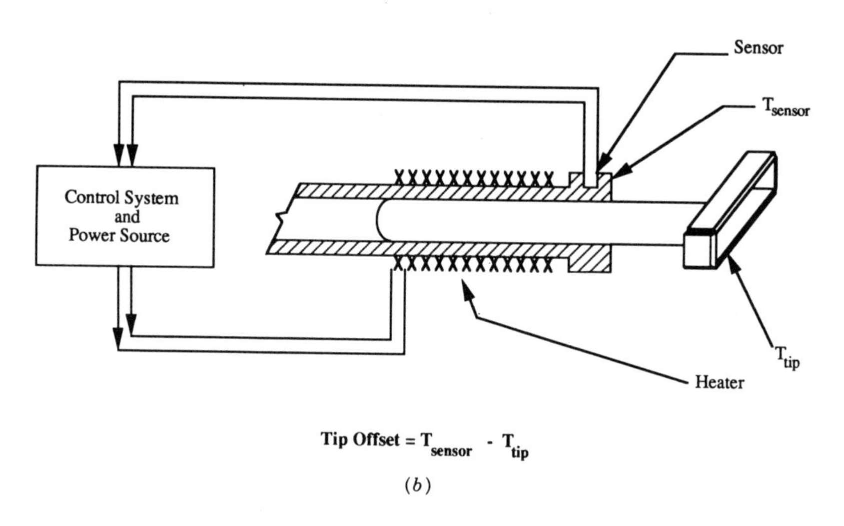

(b)

Figure 7.7: Tip offsets for through-hole and surface-mount tips

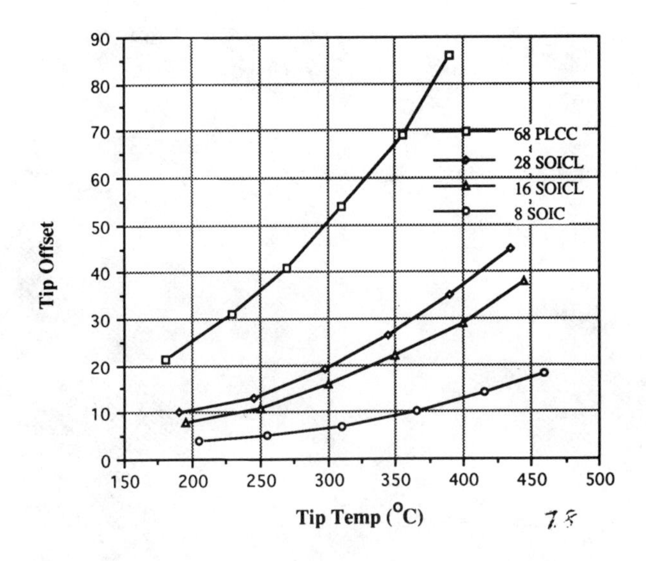

Figure 7.8: Tip offsets vs. operating tip temperature for different surface-mount tips

Removal task	Manual technique/Removal time	Semi-automated machine/Removal time
	Thermal Tweezer	**Hot Gas or IR**
PLCC 68	2-10 seconds	25-180 seconds
PLCC 84	2-10 seconds	30-180 seconds
PLCC 100	2-10 seconds	35-100 seconds
	Bar Heater with Vacuum Pick	**Hot Gas, IR or Hor Bar**
FP 68	2-5 seconds	20-120 seconds
PQFP 160	2-5 seconds	20-180 seconds
FP 208	2-5 seconds	20-240 seconds
	Soldering Iron with Bar Reflow Tips	**Hot Gas or IR**
SOICL 24	1-3 seconds	10-15 seconds
CHIP	1-3 seconds	3-10 seconds

Table 7.2: Typical removal times for various electronic components

Figure 7.9: Thermal tweezer removal of a 68-pin PLCC

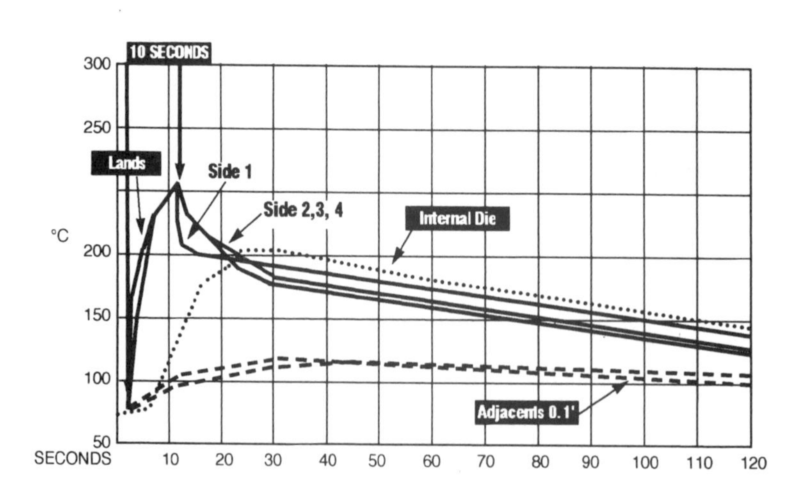

Figure 7.10: Thermal profile of thermal tweezer removal of 68-pin PLCC

Figure 7.11: Thermal pick removal of flat pack

Flat-pack/PQFP removals

Flat-pack leads are not rigidly attached to the component body as they are in a PLCC, so a thermal tweezer is not always useful for their removal. A heating block with a built-in vacuum pick up (thermal pick), such as shown in Figure 7.11, is a better choice. The thermal pick handpiece provides both heating and lifting, again permitting one-hand removals. It is usable on all forms of extended lead devices, as long as a suitably sized tip is available.

Targeting the heat at the solder joints usually results in rapid removals (see Table 7.2), with tip temperatures generally below 300°C. The thermal profile for a typical flat-pack operation is given in Figure 7.12 and shows quick uniform heat-up of the leads, low adjacent component heating, and low internal die temperatures.

SOIC/TSOP/small flat-pack removals

These popular surface-mount components are readily removed by using tips inserted into a soldering iron. Because the parts are small, the surface tension of the solder holds the component to the tip when it is lifted. Removal times are fast (Table 7.2) and thermally safe (Figure 7.13). As an alternate method, two-sided tips in a thermal tweezer may be used to remove the same components. This technique yields similar safety and speed.

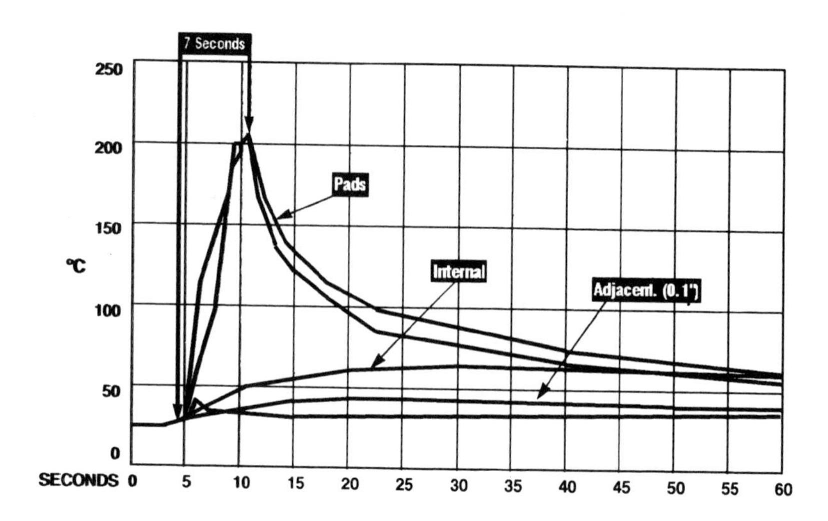

Figure 7.12: Thermal profile of a thermal pick removal of a flat pack

SIMMS module removals

The removal process for these parts is similar to that for SOICs. The main difficulty is that the parts are often tightly spaced and require thin tips for removal. For SIMMS removals, a variety of special tips that fit either a soldering iron or thermal tweezer handpiece are available.

Chip and SOT components

A variety of tips are available for soldering irons or thermal tweezers (Figure 7.15) that rapidly reflow and lift chip resistors, capacitors, MELFs, and small-outline transistor (SOT) components. Again, the surface tension of the solder provides the lifting action for these light parts. The main safety consideration occurs when removing ceramic-bodied chip components that must be reused. In this case, the assembly must be preheated, as discussed in Section 7.3.4, prior to removal of the part.

Connector and socket removals

Many two-sided thermal tweezer tips are now available for removing surface-mount connectors. The main difficulty is that the plastic used in the connector

Figure 7.13: Removal of SOIC and chip components using a soldering iron chip tip

body is easily damaged by the heat. To prevent this, a shroud of thin steel is placed over the connector body during removal. Figure 7.16 shows a typical removal. Surface-mount sockets are often removed using a socket removal tip so that the part may be directly connected to the assembly. This tip supplies heat through the leads to melt the solder joints. The tight fit of the tip into the socket then provides the lifting action to remove the part. Figure 7.18 shows a socket removal application.

7.5.2 Component and land preparation

Once the component has been removed, the excess solder should be cleaned from the surface-mount lands. Two techniques may be used to accomplish this task. The first uses solder braid or wick. A soldering iron is used to

Figure 7.14: Removal of SOIC and chip components using a soldering iron SOIC tip

heat the braid as it lays against the lands and then moves it along drawing all excess solder into the braid. This technique works well, but must be used with care. High temperatures and pressures can lift delicate surface-mount lands and insufficient soldering temperatures can cause the braid to adhere to the lands, again causing damage. This technique should be tested on scrap PCBs to establish the correct temperature.

A recent alternative to solder braid is a desoldering handpiece with a special surface cleaning tip that heats the solder and uses a continuous vacuum action to remove it from the lands. This technique is called "Flo-desoldering". Depending upon the type of cleaning involved both solder braid and Flo-desoldering techniques should be investigated and the more effective method selected. Figure 7.19 shows both cleaning methods.

It is often desirable to clean components after removal so they may be reinstalled, or to test the solderability of new components prior to installation. This technique is best done by applying flux to the component leads and then

Figure 7.15: Removal of chip components with thermal tweezer

dipping the leads into a solder pot to clean and re-tin them. Lands may be re-tinned by fluxing them and then drawing a hot soldering tip with solder across them. Temperature is critical; tinning should be done at the lowest possible temperatures. Typically, temperatures below 300°C should be used for tinning regular surface-mount lands, and even lower temperatures are desirable for fine-pitch lands. High temperatures can cause solder bridging between the lands or can damage the substrate.

Figure 7.16: Removal of surface-mount connector with thermal tweezer

Figure 7.17: Removal of surface-mount connector with thermal tweezer

7.5.3 Manual installation techniques

Once a component has been removed safely and the pad area cleaned of excess solder, the assembly is ready for new component installation. This section considers two manual techniques to install surface-mount components: conductive (with a soldering iron) and convective (with a hot-gas handpiece).

Single-point soldering

Perhaps the most obvious way to install a surface-mount component is to use a fine-point soldering-iron tip to solder each lead separately. This approach is

Figure 7.18: Removal of surface-mount socket with soldering iron tip

illustrated in Figure 7.21. Tips for this type of installation are usually conical or chisel-shaped, with front dimensions of 0.8 or 0.4 mm (1/32 or 1/64 in.). The tips are sometimes bent to permit better access into tight areas of a circuit assembly. Single-point soldering can accomplish the installation task but requires a lot of time, often 5 to 20 min. on the largest-size surface-mount components.

Bar-reflow soldering

Other hand-soldering techniques are faster than single-point soldering and yield equally good or better results. One technique uses a bar-reflow tip, as shown in Figure 7.22. For this installation, the part is first aligned and tacked in place at opposite corners. The leads are fluxed to aid solder flow, and a piece of solder wire is laid against the leads. A blade tip, such as shown in the figure, is drawn across the leads, heating and soldering each lead as it passes.

Another alternative is to draw a tip with a reservoir of solder across the fluxed leads. The tip deposits sufficient solder on each lead. The techniques have been investigated by several companies that have been satisfied with the results and attest to the reliability of the resultant solder joints. With practice, a large component with over a hundred leads can be installed in less than a minute using this approach. Temperature is again important; tip temperatures of about 300°C are used for these installations. The temperature may be adjusted slightly higher or lower to suit the speed at which the operator

Figure 7.19: Surface-mount land preparation using solder braid and Flo-desolder extractor tip

draws the solder tip across the part. As finer pitch parts are soldered, lower tip temperatures yield better results.

Hot-gas installation

In addition to conductive installations using a soldering iron, a small hot-gas jet may be used as the source of heat. Hot gas is not recommended for manual removals because of the difficulty of focusing the air flow on four sides simultaneously without overheating adjacent components. For installation, only one or two sides are reflowed at a time, allowing the operator control of the process. Furthermore, the operator uses the reflow of solder wire or solder cream to determine the rate of heat flow.

Typically, when solder creams (pastes) are used, a bead is applied to all the leads with a pneumatic dispenser. The process must be tested to determine the proper amount of cream to dispense; once this is established, the operator should be able to repeat the action. Installation techniques differ only because of the size and style of component. Some guidelines are listed below.

- **Chip/SOT installations.** Dispense the solder cream as a dot on each pad. Press the part into the solder cream, then reflow with hot gas. The gas-flow rate must be very low or the part has to be held in place to avoid being blown away as the solder begins to reflow. To reflow solder cream, start slowly until the volatiles in the cream have been dried. The cream

Figure 7.20: Surface-mount land preparation using solder braid and Flo-desolder extractor tip

changes character and seems to solidify. At this point, increase the heat until final reflow occurs. Sudden application of too much heat causes the formation of solder balls. A chip installation should take about ten seconds.

- **SOIC/flat-pack installations.** Dispense the solder cream as a continuous bead across the lands where the solder joints will form. Align and place the components into the cream. Reflow using hot gas, as indicated above. Higher gas-flow rates are necessary because of the increased size and thermal mass of the part. Again preheat the cream to achieve a dry-off prior to a final reflow. Reflow one side at a time. Total time for installation is about a minute. A picture of an SOIC installation is given in Figure 7.23, and the thermal data are shown in Figure 7.24. The heat-up to solder melt is not uniform in the figure because the tip was scanned across all of the leads and lands.

- **PLCC installations.** For this part, tack solder one lead at opposite corners to stabilize the part. Next, apply solder cream at the lead/land interface. Reflow one side at a time, starting with an untacked side. Preheat first, and then achieve final reflow. Installation for a 68-pin PLCC takes about 1 min.

Figure 7.21: Fine-point soldering tip installation of flat pack

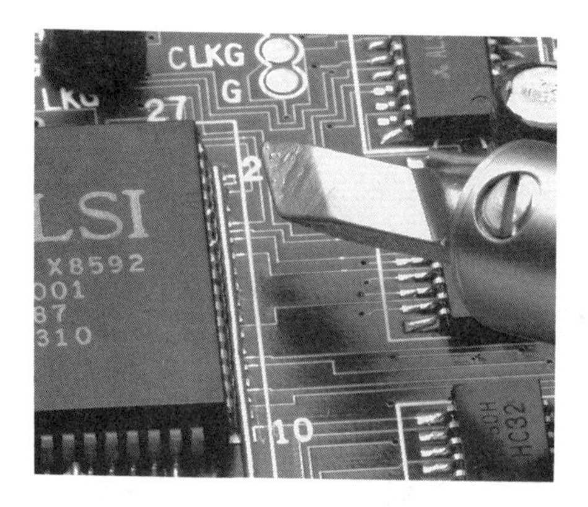

Figure 7.22: Bar-reflow installation of PLCC

Figure 7.23: Hot-jet installation of an SOIC

- **Fine-pitch installations.** Installation becomes more sensitive to the amount of solder cream being dispensed as lead spacings below 25 mils are encountered, and the composition of the solder cream also becomes more critical. These problems parallel those found in production. Operators may find better luck with the soldering-iron techniques discussed earlier when installing fine-pitch components.

- **Hot-gas installations with wire solder.** Hot-gas heating may also be used with wire solder. The area to be soldered is heated with the hot-gas tool and the solder wire is drawn across the leads.

Some final precautions on the manual use of hot-gas and solder cream are appropriate. Solder cream consistency is very important to the process. Extreme slump or liquidity in the solder cream suggests that it may be old or may not have been stored properly. Attempting reflow of bad solder cream is a frustrating experience. The solder cream does not dry off properly and solder ball formation may be severe. Another condition affecting this action is the amount of moisture in the PC assembly. PCBs with excessive moisture should

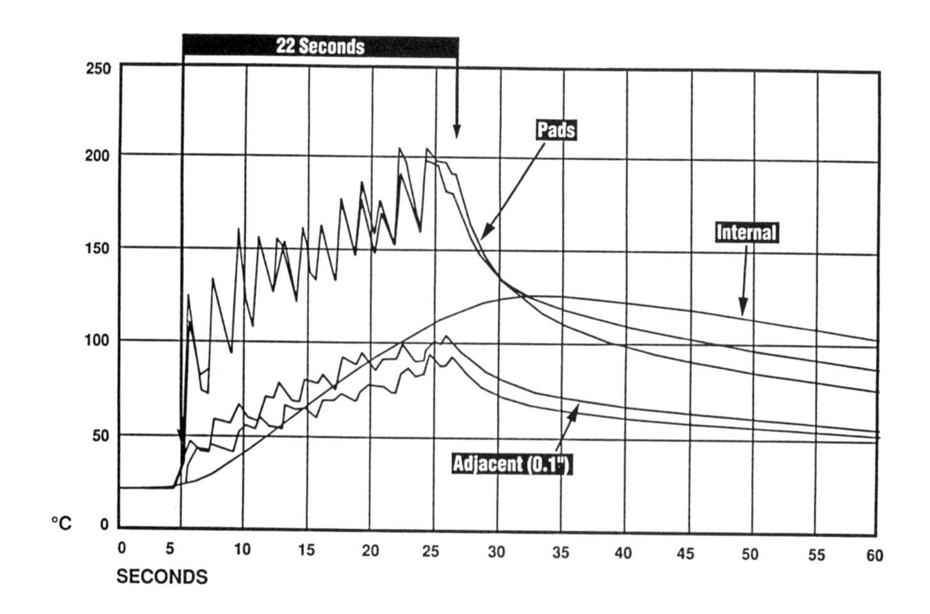

Figure 7.24: Thermal profile of hot-jet installation of an SOIC

be baked dry before rework is attempted. The excess moisture will be converted to water vapor during installation and may blister the circuit assembly.

Pulse-heat hot-bar installation

Although not as common as the other techniques, hand-held hot bar tools may also be used for surface-mount installations. The pulse-heat tips are usually not tinned, so they will not adhere to solder. For an installation, the hot-bar tip is applied cold to the leads and lands that have been previously tinned, as shown in Figure 7.25. When power is applied, the tip quickly heats up until solder melt occurs. After full reflow, the power is shut off and the tip quickly cools while the leads are stabilized.

7.5.4 Semi-automated installation and removal

Originally, nearly all surface-mount rework was accomplished by using semi-automated equipment. Many of the manual techniques described above have evolved fully within the last five years while hot-gas reflow machines have been around for over ten years. Individual machines may vary greatly in specific characteristics and the methods used to achieve a good process.

Figure 7.25: Hot-bar installation of flat pack

Hot-gas machines

Hot-gas machines, such as the one shown in Figure 7.26, were the earliest semi-automated repair systems developed for removing and installing surface-mount components. This approach was taken because of the perceived difficulty of placing and reflowing all sides of a surface-mount component simultaneously. Section 7.2.3 gives the general thermal safety guidelines the hot-gas equipment must meet. Other desirable features of hot-gas machines are summarized below:

- **Hot-gas source.** Gas-heating mechanisms differ greatly between machines, affecting reflow times for components, and equipment should be checked against the heaviest workpieces and largest components anticipated to ensure a proper reflow. In addition, the heat source should be checked for repetitive operation to ensure that it can handle a continuous stream of component installations or removals. Finally, some hot-gas systems only run from a compressed gas supply; others have internal blower supplies. This feature may be important in mobile-repair applications.

  The most significant concern in achieving correct hot-gas heating is to focus the location of the gas delivery. A reasonably rapid reflow should occur, with the heat directed primarily at the leads of the component being installed or removed. A thermal profile of a hot-gas installation that satisfies the thermal safety criteria described in Section 7.2.3 is given in Figure 7.27. Machines that saturate a large area with hot gas do not produce the proper thermal isolation for adjacent components and should not be used. There are several methods employed in better machines

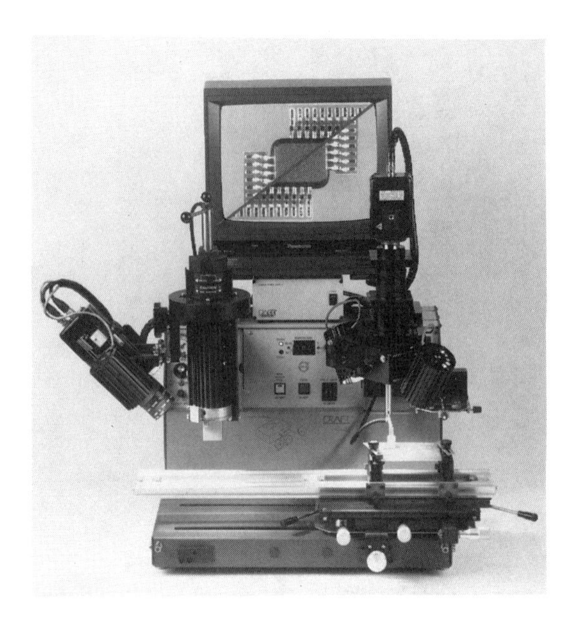

Figure 7.26: Hot-gas reflow machine

to properly focus the hot gas, including thin-wall nozzles that surround the component, cool air circulation at adjacent areas, and small scanning heaters which direct tightly focused air streams at component leads. Again, each system must be evaluated to ensure it meets the user's specific needs and the thermal guidelines in Section 7.2.3.

- **Work handling.** Most hot-gas machines have work handlers that can accommodate a wide variety of PC assemblies. These stations should provide at least the following features:
  - mounting for a wide range of different size PC assemblies;
  - coarse and fine adjustment in X, Y, and θ (rotational) directions to allow alignment of component leads to corresponding lands;
  - vacuum pick to lift parts during removal operations;
  - fixturing to allow rapid repetitive repair of similar PCBs;
  - θ shearing to aid in the removal of bonded components; and

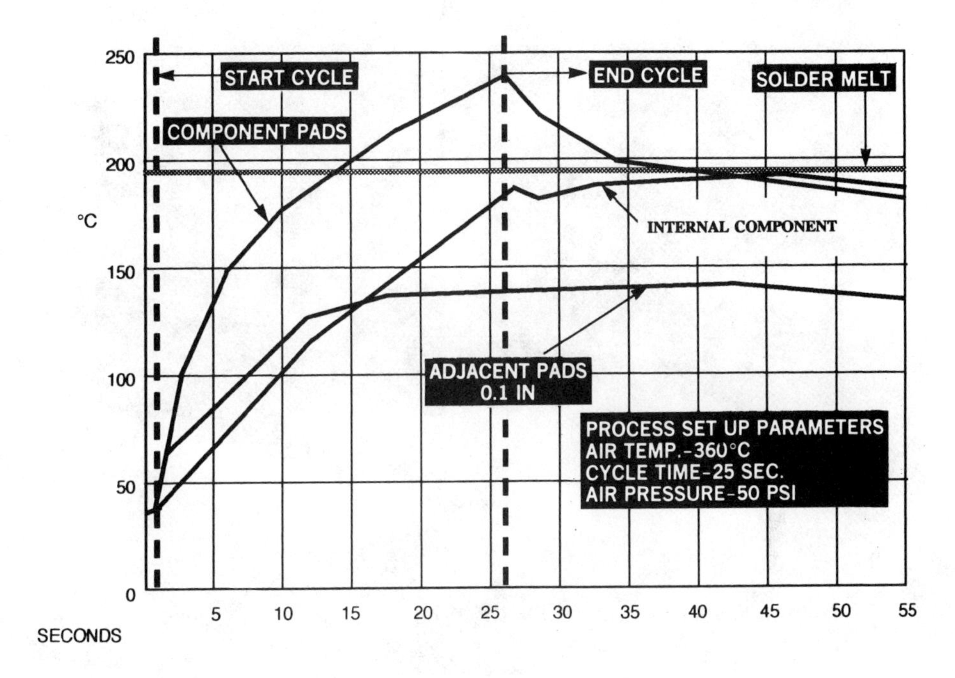

Figure 7.27: Thermal profile of hot-gas machine installation of a flat pack

  - component handling to assist in the location and placement of components.

- **Vision/lighting.** Vision systems assist in component placement, component alignment, and monitoring of the repair process. With fine-pitch components, greater demands are placed on vision systems. Split-image optics are used in some systems to allow placement of high-lead-count fine-pitch parts. The split image compresses the part so only a few leads at opposite corners are viewed at high magnification. This image allows high resolution of a few leads for placement accuracy. Figure 7.28 illustrates a split-image component view.

- **Cycle control.** When a hot-gas reflow station is set up for a specific component installation or removal, the rate of gas flow, gas temperature, and reflow time must be established and maintained. The setup is handled in a variety of ways by different equipment manufacturers. Some systems ignore setup, relying on the operator to determine the correct control settings on a trial-and-error basis. Other machines are completely manual. Some provide a process chart that gives setup parameters for different

Figure 7.28: Conventional and split-image view of 208 pin fine pitch component

component/substrate combinations. This chart usually gives the operator the correct information immediately or within one iteration. Finally, some machines provide thermal profiling as an aid to setting up the process. These cycle-control features are important because too much heat damages the substrate. Once the equipment is properly set up, the better machines automatically repeat the cycle. There is a lot of variation in this feature. Simple machines are mostly manual and rely on the operator to start and stop the process. Better hot-gas machines, however, use digital logic to ensure that cycles are accurately repeated.

Hot-bar machines

Hot-bar reflow stations have become popular in the last few years for the installation of fine-pitch components. They have not been used as general-purpose rework machines because they are usually not designed to remove parts. They are included here only because they fit the semi-automated category and can be used to reinstall these parts in a repair process.

The hot-bar machine is a sophisticated version of the concept discussed in Section 7.5.4. It normally uses thermocouple sensor feedback to control the temperature of each hot-bar heater (thermode). The heaters typically ramp up to a predetermined temperature, hold it for a fixed period of time, and then cool down on a preset ramp. The cycle time for the installation of a component is typically fast (< 30 sec.)

In addition to controlling the temperature profile, the hot-bar machines must control the pressure of the thermodes against the leads. With too much pressure, the leads are actually welded, rather than soldered, to the lands. Although the welded leads still have good electrical continuity, they are nearly impossible to repair.

Hot-bar equipment must have many of the same work handling and vision requirements of hot-gas equipment to function effectively. Because the equipment is used in a production mode, many machines have been highly automated for specific installations.

Infrared (IR) stations

In recent years, IR heating has become popular for component installations and removals. The first equipment, however, lacked good focusing and often overheated assemblies and adjacent components. More recently, focused IR equipment has been available as an alternative to hot gas. Although proponents may argue the virtues of IR versus hot gas, the better-designed versions of either machine should allow good component installations and removals. Again, the discussions of thermal safety, work handling, cycle control, and vision in Sections 7.2.3 and 7.5.4 are also applicable to IR equipment; better machines address all these concerns.

Laser

A couple of rework stations have been developed that use a laser as a heat source. Like the hot-bar machines, laser stations have been designed for production installation of fine-pitch devices. The use of laser techniques may increase in the future if leads continue to get smaller and more delicate.

Hot plate

In specific situations, a hot plate has been used to bring circuit assemblies to reflow temperatures for component installations or removals. This is a favorite salvage technique for SIMMS module boards. The danger of the hot-plate technique for general repairs is that portions of the substrate may be undesirably overheated to achieve reflow of the surface-mounted components. The hot-plate also heats many components to solder reflow at one time, and therefore does not provide suitable thermal isolation for adjacent components (see Section 7.2.3). This technique should be used with caution.

7.5.5 Pre-heating systems

The concept of preheating, discussed earlier with regard to ceramic components, is also used by many of the semi-automated machines as a part of their overall process. Because many of the machine processes take several minutes (see Table 7.2), the need to balance the expansion coefficient of PC assemblies and components may become important. Preheating systems usually use low temperature hot gas or conductive hot plates to heat the substrates to temperatures of 100° to 125°C. If hot gas is used, thermal conditions on the underside of the circuit board must be monitored, especially with a double-sided assembly. Users should make sure that solder joints are not heated above 150°C as a part of the process setup. Some of the repair machines discussed in this section offer preheating as a standard or optional feature.

7.5.6 Fully automated repair

To date, there have been limited attempts at making fully automated repair equipment. Typically, automated equipment has been manufactured by a specific company for the repair of one or more of its high-volume assemblies, rather than for general use. Because of the variety of repair problems typically encountered, most repair equipment is partially automated at best. If greater standardization occurs, more highly automated repair systems may become more common.

7.6 Other rework and repair tasks

In the previous sections, component installations and removals (the most common form of manual assembly or repair action performed with regard to printed circuit assemblies) have been discussed. As discussed in Section 7.2.2, modification or refurbishment of older assemblies creates a need for other repair actions. There are more than seventy different types of manual assembly and repair actions that can be defined (many of these are listed in Tables 7.3 through 7.7). A few of the more important actions are covered in greater detail in this section.

COMPONENT	METHOD						
	Soldering Iron	Hot Gas	Solder Extractor	Thermal Tweezer	Thermal Pick	Pulse Heat	Solder Pot
THROUGH-HOLE MOUNT							
Resistors	PI	SI	PR				
Capacitors	PI	SI	PR				
Inductors	PI	SI	PR				
Transistors	PI	SI	PR				
DIPs	PI	SI	PR	SR			
Pin Grids	PI	SI	PR				
Connectors	PI	SI	PR			PI	
Sockets	PI	SI	PR				
Cables	PI	SI					
Circuit Tracks		SI				PI	
Terminals	PI		PR			PI	
Other Parts	PI	SI	PR			SI	

PI = Primary installation
PR = Primary removal
PI/R = Primary installation/removal
SI = Secondary installation
SR = Secondary removal
SI/R = Secondary insallation/removal
PH = Preheat

Table 7.3: Manual installation/removal functional matrix for through-hole packages

COMPONENT	METHOD						
	Soldering Iron	Hot Gas	Solder Extractor	Thermal Tweezer	Thermal Pick	Pulse Heat	Solder Pot
SURFACE MOUNT							
PLCCs/LCCs	PI	PI		PR		PI/R	
Flat Packs/PQFPs	PI	PI		SR	PR		
Chips/SOTs/MELFs	PI/R	PI/R		PR			
SOICs/TSOPs	PI/R	PI		PR			
SOJs/SIMMs	PI/R	PI		PR			
Connectors	PI	PI		PR			
Sockets	PI/R	PI					
Fine Pitch	PI	PI		PR			
Other Parts	PI	PI					
Ceramic Parts	PI			PR			
Bonded Parts	PI			PR			
Land Preparation	PR		PR				
Component Preparation	PI		PI				PI

PI = Primary installation
PR = Primary removal
PI/R = Primary installation/removal
SI = Secondary installation
SR = Secondary removal
SI/R = Secondary insallation/removal
PH = Preheat

Table 7.4: Manual installation/removal functional matrix for surface-mount packages

COMPONENT	METHOD					
	Hot Gas	Hot Bar	Hot Plate	Infrared	Laser	Solder Fountain
THROUGH-HOLE MOUNT						
Resistors						
Capacitors						
Inductors						
Transistors						PI/R
DIPs	PI					PI/R
Pin Grids	PI					PI/R
Connectors						PI/R
Sockets						
Cables						
Circuit Tracks						
Terminals						
Other Parts						

PI = Primary installation
PR = Primary removal
PI/R = Primary installation/removal
SI = Secondary installation
SR = Secondary removal
SI/R = Secondary insallation/removal
PH = Preheat

Table 7.5: Semi-automated installation/removal functional matrix for through-hole packages

COMPONENT	METHOD					
	Hot Gas	Hot Bar	Hot Plate	Infrared	Laser	Solder Fountain
SURFACE MOUNT						
PLCCs/LCCs	PI/R			PI/R	PI/R	
Flat Packs/PQFPs	PI/R	PI		PI/R	PI/R	
Chips/SOTs/MELFs						
SOICs/TSOPs	PI/R			PI/R	PI/R	
SOJs/SIMMs	PI/R		SR	PI/R	PI/R	
Connectors						
Sockets						
Fine Pitch	PI/R	PI		PI/R	PI/R	
Other Parts						
Ceramic Parts	PI/R		PH	PI/R	PI/R	
Bonded Parts	PI/R			PI/R	PI/R	
Land Preparation			PR			
Component Preparation			PI			

PI = Primary installation
PR = Primary removal
PI/R = Primary installation/removal

SI = Secondary installation
SR = Secondary removal
SI/R = Secondary insallation/removal

PH = Preheat

Table 7.6: Semi-automated installation/removal functional matrix for surface-mount packages

REPAIR TYPE	EQUIPMENT NORMALLY USED								
	Pulse Heat	Machine Tool	Pulse Plate	Solder Tool	Fused Eyelet	Special Plater	Spot Cleaner	Vacuum Cleaner	Optical Aids
REMOVAL TASK									
Damaged Track		X							
Plated Through-Hole		X							
Edge Connector		X							
Rigid Coating	X	X							
Flexible Coating	X								
Paralene		X							
INSTALLATION TASK									
Damaged Track	X			X					
Plated Through-Hole				X	X				
Edge Connector	X			X					
MODIFICATION TASK									
New Component	X			X					
Jumper		X							
Probing		X							
X = Normally used for this procedure									

Table 7.7: Circuit level repair matrix

REPAIR TYPE	EQUIPMENT NORMALLY USED								
	Pulse Heat	Machine Tool	Pulse Plate	Solder Tool	Fused Eyelet	Special Plater	Spot Cleaner	Vacuum Cleaner	Optical Aids
REFURBISHMENT TASK									
Cleaning Traces		X	X						
Plating Traces			X						
Plated Through-Holes						X			
SPECIALTY TASK									
Cleaning							X		
Component Handling								X	
Component Alignment								X	X
Thermal Wire									
Stripping	X								
Preheating Leads	X			X					
X = Normally used for this procedure									

Table 7.7: Circuit level repair matrix (contd.)

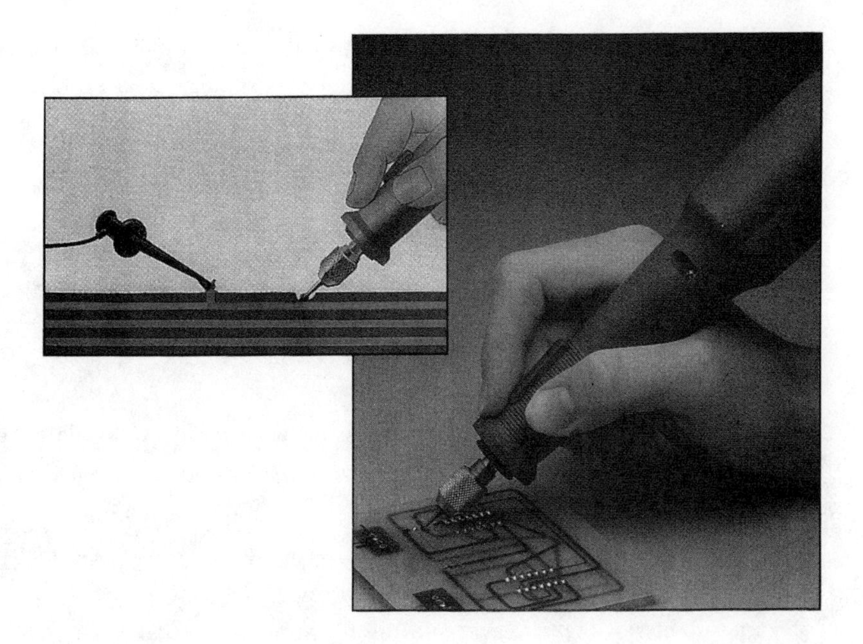

Figure 7.29: Modern machining tool for surface-mount repair, showing probe brake feature to permit intelligent machining

7.6.1 Track repair

Often, to make a circuit modification an existing track must be exposed or cut and jumpers or additional components must be added. In other situations, a circuit assembly may have a defective track that needs to be repaired. Tracks are exposed by using a machining tool to remove the covering substrate material. Nowadays, multilayer designs have made this type of repair difficult to accomplish, because the circuit change may require access to an inner layer.

Advances in machining technology have helped to overcome this problem by allowing "intelligent machining". This technique is carried out as indicated in Figure 7.29. The track that must be probed or machined is electrically activated with a low-level signal. When the machine tool touches this track, it stops immediately, preventing damage and signaling the operator. This form of intelligent probing makes track repairs and adding signal jumpers relatively easy to accomplish. Normally, track repairs are carried out using the following steps:

- **Identifying repair area.** Identification may be accomplished in a variety of ways: referring to schematics, using a bare board, using component-level or functional-test equipment, or manually probing the assembly.

- **Determining the repair action.** The form of repair action can vary, but may include adding jumpers, repairing a broken track, replacing lands, or refurbishing plated through-holes.

- **Preparing the defective circuit area.** Preparation includes machining to expose the broken track or land area, and physically removing defective track.

- **Replacing defective lands.** This task is made easier by the availability of track and pad repair kits, such as shown in Figure 7.30. In recent years, these kits have been extended for use on surface-mount assemblies. Usually the surface-mount tracks have an adhesive backing that is thermally activated to bond them in place. A variety of eyelets and funnelets have been available for many years to facilitate through-hole pad repairs. Recently, equipment for re-plating the interior of plated through-holes has also been available.

- **Completing electrical connections.** This step may include a variety of different activities. Those most commonly encountered are creating lap or butt joints to connect new tracks to existing ones. The technique is facilitated by using a hot-bar "lap-flow" tip, as indicated in Figure 7.31. This tip is applied cold to the work, heated to achieve a good solder connection, and then held in place while it rapidly cools to stabilize the new lead. It is a special form of the pulse-heated hot-bar equipment discussed in Section 7.5.4. Funnelets may be soldered in place to establish electrical continuity in a plated through-hole. An alternative to this action is hot fusing an eyelet in place, although this technique is not always considered acceptable for long-term repairs.

- **Rebuilding the PC assembly.** If the circuit board has been machined to achieve the repair, the machined areas are usually filled with an epoxy material after the repair is made. This process restores the structural integrity of the assembly. Section 7.6.3 covers plating new leads or lands.

The time to repair a circuit assembly is typically longer than the time to remove or install a component. Thus, circuit-level repairs are not

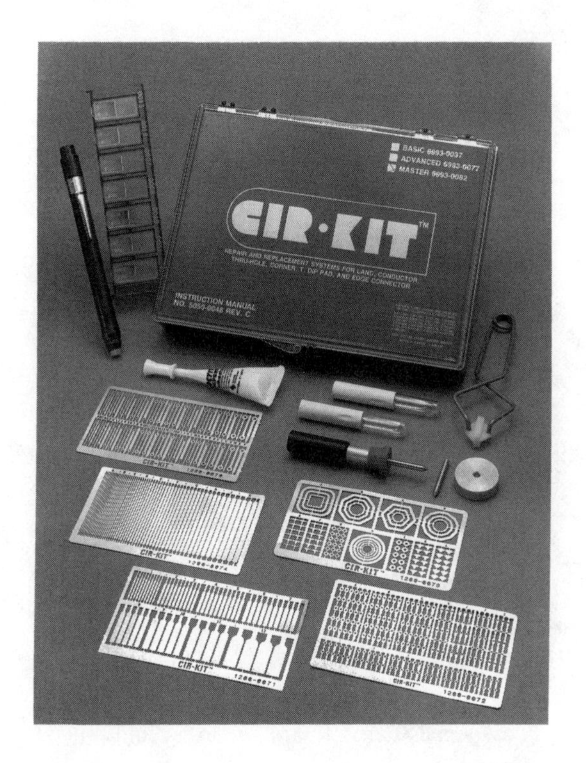

Figure 7.30: Replacement track pads, eyelets, and funnelets

normally made on low-cost assemblies for economic reasons. Circuit-level repairs are common, however, on more expensive commercial or military assemblies, where the repair cost can be justified against either the cost of scrapping an entire assembly or the loss of a critical operation.

7.6.2 Conformal-coating removal

Many military and some commercial assemblies are conformally coated to protect the circuitry from environmental contamination. When these assemblies must be repaired, it is often necessary to remove the conformal coating in the area of the repair and replace it after the repair has been made. Removal techniques depend on the coating material being used.

- **Rigid (epoxy) coatings.** These are usually abraded away from the repair area, using a machining tool such as described in Section 7.6.1 and

Figure 7.31: Lap-flow soldering of repaired circuit track

shown in Figure 7.32. A variety of special small bits and wire brushes are available for this type of work.

- **Rubberized coatings.** Depending upon consistency, these coatings may either be abrasively removed or thermally softened and removed. For thermal removal, a special thermal parting tip, such as shown in Figure 7.33, is often used. It acts much like a hot scalpel and greatly simplifies the removal task.

- **Vacuum-deposited coatings.** Paralene^R, one of the more popular coatings, is vacuum-deposited onto a circuit assembly and reportedly provides a high-quality barrier that is impervious to most chemical and other environmental hazards. This feature makes chemical removal of this coating nearly impossible. A variety of techniques are used to remove the coating as a part of a repair action, including the heating and abrasive methods already discussed. In addition, vapor-blasting techniques have been successfully used to wear away the coating. Because this type of coating is very thin, it is frequently left in place when surface-mount components are removed. The coating and the part are removed at the same time. To ensure clean removal, the coating may have to be scored with a sharp knife around the periphery of the component so that it separates cleanly.

This section provides an overview of possible coating-removal actions. Users should obtain and follow specific manufacturers' recommendations when applying or removing these conformal coatings.

Figure 7.32: Abrasive removal of rigid conformal coating

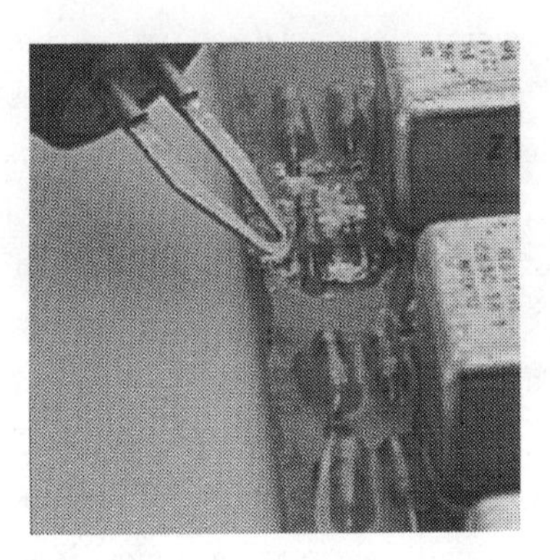

Figure 7.33: Thermal removal of conformal coating

7.6.3 Electroplating and cleaning

Electroplating is an old repair process that has been used for many years to clean and refurbish edge connectors and replate circuit tracks that have been replaced (see Section 7.6.1). The technique, often called swab plating, uses a handpiece with a conductive end covered with a cotton material; it is dipped into the plating solution and wiped across the surface of the PCB. Although many metallic materials can be plated using these techniques, the most common for PCB repairs are nickel and gold. A sodium-hydroxide solution is often used

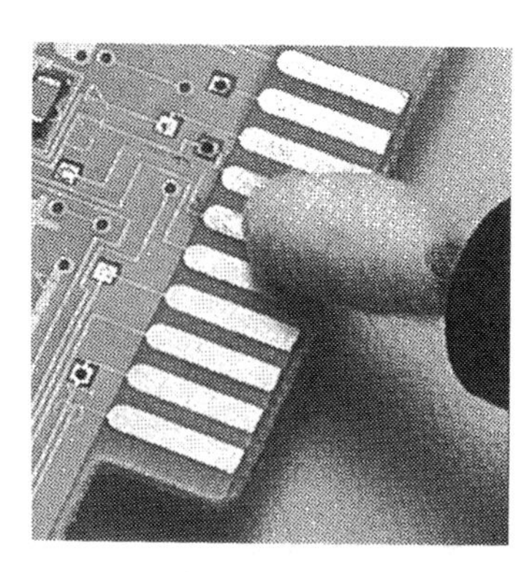

Figure 7.34: Gold plating an edge connector

for electrocleaning circuit traces to remove oxides prior to plating.

Plating supplies generally use a DC source, although swab-plating depositions using pure DC tend to vary greatly in thickness. To allow for a more uniform deposition, modern swab-plating power sources add a pulsed AC component to the plating signal. Swab-plating develops larger current densities than conventional electroplating, so depositions take place quickly. With practice, the hand plating can be controlled very well and yields acceptable quality.

The most common use of swab-plating equipment is still the refurbishment of edge connectors (see Figure 7.34). The process usually includes electrocleaning to remove contaminants and oxides from the copper substrate, a pre-plating nickel deposition, and a final gold-plating deposition. Recently, new plating equipment has been designed to allow replating of plated through-holes. The hole and plating solution are electrically energized, and the plating solution is moved back and forth through the hole.

7.6.4 Specialty soldering tasks

Many assembly and repair tasks are difficult to conduct with only a simple soldering iron. For example, the butt- or lap-joint soldering of a replaced track (discussed in Section 7.6.1) uses a pulse-heat handpiece that both stabilizes the track and reflows the solder. Another pulse-heat soldering application uses a resistance heating approach that melts the solder by passing a current through a part, such as a connector pin or wire post. Although it may not be desirable to use this approach on a circuit component, it is often useful for connector

Figure 7.35: Installing wires in a connector, using resistance soldering

and turret terminal soldering.

Figure 7.35 shows a connector-soldering application. In this instance, the connector pin is gripped by the conductive handpiece and heated quickly by passing current through it. When solder melt occurs, the wire is inserted (or removed), after which the current is shut off to let the solder solidify. Other special pulse-heat tasks, listed in Table 7.3, are carried out during manual assembly and repair as required.

7.7 The modern assembly/repair station

In the past, an assembly station consisted primarily of a soldering iron. Later, a continuous vacuum solder extractor became a part of most stations to permit through-hole repairs. With the advent of surface-mount technology, however, these hand tools alone became insufficient to handle all manual assembly and repair tasks. This section provides guidelines to aid in the selection of equipment for specific or general use in the manual or semi-automated assembly or repair of surface-mount and through-hole PCBs.

7.7.1 Repair aids

The complete outfitting of an assembly or repair workstation before considering specific equipment requires new accessories to deal with needs dictated by surface-mount assembly or repair.

Figure 7.36: Vacuum handling of delicate surface-mount components

Visual aids

Surface-mount components have more leads of finer size and pitch spacing than through-hole parts. Their installation requires visual aids to ease operator eye strain and aid the resolution ability of the naked eye. For normal surface-mount parts with lead spacings of 25 to 50 mils, many operators can align and install parts with the naked eye; however, some form of low-level (3X to 10X) magnification is usually helpful. For finer-pitch parts with lead spacings less than 25 mils, magnification is mandatory. Such visual aids ensure lead-to-land alignment and aid final inspection after assembly or repair. In some modern stations, a video camera and monitor are used in place of a microscope or lens.

Component handling

Disturbing the delicate lead structure of many modern surface-mount components may damage a lead or, at a minimum, upset lead coplanarity. These problems may be avoided by using tools suited to handle delicate parts, such as tweezers or vacuum picks (Figure 7.36). Handling most surface-mount components by hand is not recommended.

New materials

Several of the techniques discussed in earlier sections use solder creams. When these techniques are employed, manual or, preferably, automated dispensers are required.

EOS/ESD safety

All of the normal issues related to safe assembly of PCBs are important to surface-mount components. The equipment used to install and remove parts should meet the technical requirements mentioned in Section 7.4.1. In addition, most modern equipment has handles and cord assemblies that are manufactured with static-dissipative materials. Air-generating equipment should be checked to verify that it is designed to minimize the generation of a static charge. Users should verify the safety aspects of all equipment prior to use.

7.7.2 Functional equipment selection

There are many different ways to determine which new equipment is needed in a facility; one is a functional approach. A partial list of the general functions that might be required are:

- component-level functions
  - through-hole component installation
  - through-hole component removal
  - surface-mount component installation
  - surface-mount component removal
- circuit-level functions
  - machining
  - conformal-coating removal
  - track soldering
  - electroplating
- specialty functions
  - connector assembly
  - pulse-heat soldering
  - other tasks (see Table 7.3)

Determining functional needs

From the preceding list and Table 7.3, the desired functions for any work area can be determined. For example, in a repair center, one station providing component-level and circuit-level repairs might be desired, together with several stations designed only for through-hole and surface-mount component installation and removal. In a production area, more specialized applications might be desirable. For example, many areas might need only a soldering iron for some simple assembly tasks. Other areas may be dedicated to through-hole repair and need only a soldering iron and an extractor. Still other areas might handle only surface-mount assembly and repair actions and require other handpieces, or a specific machine system. A functional list targets the specific needs for each area and then allows the definition of specific equipment needs.

Planning for the future

Planning that doesn't allow for some future tasks can often quickly cost companies additional money for new equipment or the early obsolescence of recently purchased equipment. Companies should plan for present needs with as much consideration of future needs as can be estimated reasonably. On the other hand, the danger of attempting to plan so far into the future that no product will be suitable should also be avoided.

Process control, productivity, and price

Finally, the repair philosophy of a specific company must be considered in selecting the type of equipment that will yield the desired process control and safety. This philosophy differs among companies, and each company usually sets its own specific set of standards or "house rules." Whatever the company's specific requirements, they must still satisfy the thermal safety criteria of Section 7.2 if successful repairs are expected. Most companies have enjoyed the flexibility of through-hole assembly and repair equipment that can be easily set up and used in almost any area of a plant. Many larger semi-automated systems for surface-mount assembly or repair do not allow the same flexibility; rather, they become dedicated facilities in specific plant areas. The repair tasks must be brought to this area rather than being handled throughout the plant. In contrast, manual surface-mount techniques and equipment offer the same flexibility as through-hole soldering and desoldering stations.

Process control. The role of human operators in the repair process changes, depending on the assembly and repair philosophy a company follows. Some companies like the concept of a specific repair area to control the operation carefully. Others consider only partially or fully automated repair processes because of concerns about costs or a lack of quality with a manual approach. Whatever the justification, this decision dramatically affects the equipment choices a company makes. The right choice is the one that permits high-quality assemblies and repairs (excellence).

Productivity. New equipment should be evaluated to determine its suitability in specificapplications. The best gauge is often the productivity of the process, assuming safety needs have been satisfied. The faster the process, the wider the range of applications that can be considered. Slow processes might cost more than scrapping assemblies, whereas fast processes can often save companies a great deal of money. (Some guidelines on process speed for surface-mount installations are covered in Section 7.5 and summarized in Table 7.2.)

Price. Finally, once assembly and repair needs are functionally determined and process control and productivity are established, the cost to implement various functions can be determined. The only additional guideline is that products that group functions can often save bench space and money. For example, if it is desirable to have a manual bench-top station that can handle through-hole and surface-mount assembly and repair, several options are available, including a soldering iron station, a soldering extractor station, a thermal tweezer station, a hot-gas station, or a thermal pick station. A grouping of all these staions can accomplish the task, but takes up a considerable amount of bench space and has a relatively high cost. Equipment that offers several or all of the required functions in one unit takes up less bench-top space and is usually more cost-effective.

The functional approach should greatly assist individuals or companies in making educated selections of the best and most cost-effective equipment for their surface-mount or through-hole assembly and repair. All equipment must meet thermal and mechanical safety criteria (nature's rules), but the final choice must also satisfy a company's specific goals on process control, productivity, and price (house rules).

7.8 References

Abbagnaro, L. Evaluating Continuous Vacuum Desoldering Systems. PACE, Inc. *Electronics Manufacturing* (September 1988) 15-19.

Abbagnaro, L. Process Development in SMD Rework. PACE, Inc. *Circuits Manufacturing* (December 1988) 50-55.

Abbagnaro, L. Soldering and Mounting Technology. *Electronic Materials Handbook*, Volume 1 Packaging, ASM, Ohio, 1989.

Abbagnaro, L. Safe Repair of Surface-Mount Assemblies. PACE, Inc. *Electronic Packaging and Production* (November 1990) 54- 61.

Abbagnaro, L. SMD Assembly and Repair. PACE, Inc. *Surface Mount Technology* (August 1991) 28-33.

Abbagnaro, L. Rework and Repair Revisited. PACE, Inc. *Circuits Assembly* (May 1992) 26-35.

ANSI/IPC-R-700C. Suggested Guidelines for Modification, Rework and Repair of Printed Circuit Boards and Assemblies. Revision C, IPC, Lincolnwood, IL (January 1988).

Bausell, J. Focus on Soldering and Desoldering. PACE, Inc. (Electronic Servicing & Technology) (November 1989) 12-18.

Economu, M. Extra-Fine-Pitch SMT Rework. Hewlett Packard. *Circuits Assembly* (May 1992) 36-39.

Hodson, T. Solving Rework and Repair Problems. *Electronic Packaging & Production* (January 1992).

Scarlett, J. The Multilayer Printed Circuit Handbook. Scotland: Electrochemical Publications, Ltd. (1988).

Siegel, E. Where's the Heat, Part 1. PACE, Inc. *Circuits Assembly* (November 1989) 10-17.

Siegel, E. Where's the Heat, Part 2. PACE, Inc. *Circuits Assembly*, (March 1991) 58-67.

Appendix

Suppliers of Soldering Equipment

3M Company
Industrial Chemical Products Division
St. Paul, MN

Air Products and Chemicals, Inc.
Allentown, PA

Air-Vac Engineering Company, Inc.
Milford, CT

Anorad Corporation
Hauppauge, NY

Asymtek International Headquarters
Carlsbad, CA

Austin American Tech. Corp.
Austin, TX

Automated Production Concepts, Inc.
Wayne, NJ

Benchmark Industries, Inc.
Goffstown, NH

BGK Electronic Industries
Division
Minneapolis, MN

BTU International, Inc.
North Billerica, MA

Brian R. White Company Inc.
Ukiah, CA

Camelot Systems Inc.
Haverhill, MA

Centech Corporation
Minneapolis, MN

Comatel USA, Inc.
Danbury, CT

CONCEPTRONIC
Exeter, NH

Control Laser Corporation
Orlando, FL

Cooper Tools
Raleigh, NC

Corpane Industries, Inc.
Louisville, KY

Cyber Optics Corporation
Minneapolis, MN

Detrex Corporation
Equipment Division
Southfield, MI

Dialight Corporation
Manasquan, NJ

Dynapert Division
Beverly, MA

E.P.M. Handels A.G.
Switzerland

Electronics Controls Design, Inc.
Milwaukie, OR

Electrovert USA
Arlington, TX

Elvo
Division of Zumbach Electronics
Corporation
Mount Kisco, NY

EMD Associates, Inc.
Winona, MN

Exselect Engineering, Limited
Concord, ON
Canada

Farco USA
Clinton, MA

Fox Laboratories
Danbury, CT

General Microcircuits, Inc.
Moeresville, N.C.

Hexacon Electric
Roselle Park, NJ

Hollis Engineering
Nashua, NH

Hughes Aircraft
Industrial Products Division
Carlsbad, CA

Kirsten Kabeltechnik AG
4716 Welschenrohr
Switzerland

Koki Company Limited
Chao-ku, Tokyo
Japan

Light Soldering Developments
Limited
Croyden, Surrey
England

Manufacturing Technology
San Jose, CA

MCT/Browne, ITS
Vernon, CT

Mechanization Associates
Mountain View, CA

Merlin Machinery, Inc.
Somerville, MA

Mitsubishi International
Corporation
Palo Alto, CA

Nihon Den-Netsu Keiki
Company Ltd.
Ohta-ku, Tokyo
Japan

O.K. Industries Inc.
Yonkers, NY

Oal Associates, Inc.
Escondido, CA

PACE, Inc.
Laurel, MD

Philips
Industrial Automation Division
Norcross, GA

Plato Products, Inc.
Glendora, CA

NAS Electronics
Hackensack, NJ

Prodomax Industrial Automation
Barrie, ON
Canada

Research, Inc.
Assembly Automation Division
Minneapolis, MN

Robotic Process Systems, Inc.
Simi Valley, CA

Roken Electronics (Europe) Limited
Andover, Hants
England

Schunk Automation Systems
Old Saybrook, CT

Semiconductor Equipment Company
Moorpark, CA

Sensbey, Inc.
Burlingame, CA

Sikama International
Santa Barbara, CA

Streckfuss USA, Inc.
Irving, TX

Tamura Seisukusho Company Limited
Nerima-ku, Tokyo
Japan

TDK Corporation of America
Mount Prospect, IL

Technical Devices Company
Culver City, CA

The Christopher Group
Santa Ana, CA

The John Treiber Company
Fountain Valley, CA

U.S. Laser Corporation
Wyckoff, NJ

Union Carbide Industrial Gases, Inc.
Danbury, CT

Unit Design, Inc.
Orange, CA

Universal Instruments Company
Binghamton, NY

Vitronics Corporation
New Market, NH

WENESCO
Chicago, IL

Glossary of Soldering Terms

Absorbed contaminant - A contaminant attracted to the surface of a material held captive in the form of a gas, vapor, or condensate.

Accelerated aging - The means whereby the deterioration encountered in natural aging is artificially reproduced and hastened.

Acceptable quality level (AQL) - The maximum number of defects per one-hundred units that can be considered satisfactory as a process average.

Acceptance tests - Tests deemed necessary to determine the acceptability of a product and as agreed to by both purchaser and vendor.

Access hole - A hole or series of holes in successive layers of a multilayer board, providing access to the surface of a land on one of the layers of the board.

Accuracy - The deviation of the measured or observed value from the accepted reference.

Acid flux - An aqueous solution of an acid and an inorganic, organic, or water-soluble flux. (See also "Inorganic flux" and "Water-soluble organic flux.")

Acid-core solder - Wire solder with a self-contained acid flux.

Activated rosin flux - A mixture of rosin and small amounts of organic-halide or organic-acid activators. (See also "Synthetic activated flux.")

Activating - A treatment that renders nonconductive material receptive to electrodeless deposition.

Activating layer - A layer of material that renders a nonconductive material receptive to electroless deposition.

Activator - A substance that improves the ability of a flux to remove surface oxides from the surfaces being joined.

Adhesion layer - The metal layer that adheres a barrier metal to a metal land on the surface of an integrated circuit.

Advanced statistical method - A statistical process analysis and control technique that is more sophisticated but less widely applicable than basic statistical methods.

Aging - The change of a property (e.g., solderability) with time. (See also "Accelerated aging.")

Air pollution - Contamination of the atmosphere with toxic or otherwise harmful substances.

Air contamination - See "Air pollution."

Alignment mark - A stylized pattern selectively positioned on a base material to assist in alignment.

Ambient - The surrounding environment in contact with a specified system or component.

Anchoring spur - An extension of a land on a flexible printed board extending beneath the cover lay to hold the land to the base material.

Annular ring - The portion of conductive material completely surrounding a hole.

Aqueous flux - See "Water-soluble organic flux."

Area array tape automated bonding - Tape-automated bonding where some carrier tape terminations are made to lands within the perimeter of the die.

Aspect ratio (hole) - The ratio of the length or depth of a hole to its preplated diameter.

Assembled board - See "Assembly."

Assembly - A number of parts, subassemblies, or combinations thereof joined together. (Note: This term can be used in conjunction with other terms e.g., "Printed wiring board assembly.")

Automated component insertion - The act of assembling discrete components to printed boards by means of electronically controlled equipment.

Axial lead - Lead wire extending from a component or module body along its longitudinal axis.

Backlighting - Viewing or photographing by placing an object between a light source and the eye or recording medium.

Backpanel - See "Backplane."

Backplane - An interconnection device used to provide point-to-point electrical interconnections (usually a printed board with discrete wiring terminals on one side and connector receptacles on the other) (See also "Mother board.")

Bare board - An unassembled (unpopulated) printed board.

Base solderability - The ease with which a metal or alloy can be wetted by solder under minimum realistic conditions.
Basic statistical method - The application of a theory of variation through the use of basic problem-solving techniques and statistical process control. (This includes control and capability analysis for both variables and attributes data.)
Basic wettability - The ease with which a specific metal or alloy can be wetted by solder.
Binomial distribution - A discrete probability distribution that, with certain assumptions, describes the way that attributes (proportions) vary.
Blind via - A via extending only to one surface of a printed board.
Board thickness - The overall thickness of the base material and all conductive materials deposited thereon.
Board - See "Printed board" and "Multilayer printed board".
Boss - See "Land."
Bump - A raised metal feature on a die land or tape carrier tape that facilitates inner-lead bonding.
Bumped tape - Carrier tape with raised metal features that facilitate inner-lead bonding.
Buried via - A via that does not extend to the surface of a printed board.
Burn-in - The process of electrically stressing a device temperature for a sufficient length of time to cause the failure of marginal devices.
Burn-in, dynamic - Burn-in at high temperatures that simulates the effects of actual or simulated operating conditions.
Burn-in, static - Burn-in at high temperatures with unvarying voltage, either forward or reverse bias.

Card - See "Printed board."
Check list - A list of the specified criteria that may be included in an audit or inspection.
Chip - See "Die"
Chip carrier - A low profile, usually square, surface-mount component semiconductor package, leaded or leadless, whose die-cavity or die-mounting area is a large fraction of the package size and whose external connections are usually located on all four sides of the package.
Chip-on-board (COB) - A printed board assembly technology that uses un-

packaged semiconductor die.

Circuit card - See "Printed board."

Circuitry layer - A layer of a printed board containing conductors, including ground and voltage planes.

Clinched lead - A component lead which is inserted through a hole in a printed board and which is then formed to retain the component in place and to make metal-to-metal contact with a land prior to soldering. (See also "Partially clinched lead.")

Clinched-wire interfacial connection - See "Clinched-wire through connection."

Clinched-wire through connection - A connection made by a bare wire that has been passed through a hole in a printed board and subsequently formed (clinched) and soldered to the conductive pattern on each side of the board.

Coefficient of thermal expansion (CTE) - The linear dimensional change of a material per unit change in temperature. (See also "Thermal expansion mismatch.")

Coined lead - The end of a round lead that has been formed to have parallel surfaces that approximate the shape of a ribbon lead.

Cold solder connection - A solder connection that exhibits poor wetting and is characterized by a grayish, porous appearance due to excessive impurities in the solder, inadequate cleaning prior to soldering, and/or the insufficient application of heat during the soldering process.

Component - An individual part or combination of parts that, when together, perform a design function(s). (See also "Discrete component.")

Component hole - A hole used for attaching and electrically connecting component terminations, including pins and wires, to a printed board.

Component lead - The solid or stranded wire or formed conductor that extends from a component to serve as a mechanical or electrical connector, or both.

Component mounting - The act of attaching components to a printed board, the manner in which they are attached, or both.

Component mounting orientation - The direction in which the components on a printed board or other assembly are lined up electrically with respect to the polarity of polarized components, with respect to one another, and/or with respect to the board ouline.

Component pin - A component lead that is not readily formable without being damaged. (See also "Component lead.")

Computer-aided design (CAD) - The interactive use of human operators and data-manipulating computer systems, programs, and procedures in the design process.

Computer-aided engineering (CAE) - The interactive use of human operators and data-manipulating computer systems, programs, and procedures in an engineering process.

Computer-aided manufacturing (CAM) - The interactive use of human operators and data-manipulating computer systems, programs, and procedures in various phases of a manufacturing process.

Computer download - Transferring computer programs or data from a computer to a lower-level computer.

Confidence interval - The determination, with a specified degree of confidence, of whether a particular characteristic is within ascertained limits for a population.

Conformal coating - An insulating protective covering that conforms to the configuration of the objects coated when applied to a completed printed board assembly.

Connector - A device used to provide mechanical connect/disconnect service for electrical terminations.

Connector contact - The conducting member of a connecting device that provides a separable connection.

Contact angle - The angle of a solder fillet enclosed between a plane tangent to the solder/basis-metal surface and a plane tangent to the solder/air interface.

Contact area - The common area between a conductor and a connector through which the flow of electricity takes place.

Controlled collapse soldering - Creating termination by controlling the height of solder balls in a flip-chip assembly operation.

Coplanar leads - The flat beam leads of a component package that have been formed so that they can simultaneously contact one plane of a base material.

Copper-mirror test - A test of the corrosivity of a flux on a copper film vacuum-deposited on a glass plate.

Corrosion (chemical/electrolytic) - The attack of chemicals, flux, and flux residues on base metals.

Corrosive flux - Flux that contains levels of halides, amines, or organic acids that cause copper corrosion.

Cover coat - See "Cover lay."

Cover lay - The layer of insulating material applied over a conductive pattern on the outer surface of a printed board.

Cover layer - See "Cover lay."

Creep - Time-dependent strain occurring under stress.

Critical defect - Any anomaly specified as being unacceptable.

Critical operation - A part of a total process that has a significant impact on the characteristics of the completed product.

Cup solder terminal - A cylindrical solder terminal with a hollow opening into which one or more wires are placed prior to soldering.

Cycle time - The reciprocal of the maximum throughput time.

Defect - Any nonconformance to specified requirements by a unit or product.

Defect identification - The provision for locating a detected anomaly.

Degradation - A decrease in the performance characteristics or service life of a product.

Dendritic growth - Metallic filaments that grow between conductors in the presence of condensed moisture and an electric bias. (See also "Whisker.")

Device - An individual electrical circuit element that cannot be further reduced without destroying its stated function.

Dewetting - A condition that results when molten solder coats a surface and then recedes, leaving irregularly shaped mounds of solder separated by areas covered with a thin film of solder.

Dice - More than one die.

Die - The uncased and normally leadless form of an electronic component that may be either active or passive, discrete or integrated.

Digitizing - Converting feature locations on a flat plane to digital representation in X-Y coordinates.

Dip soldering - Making of multiple soldered terminations simultaneously by bringing the solder side of a printed board with through-hole mounted components into contact with the surface of a static pool of molten solder. (See also "Drag soldering.")

Discrete component - A separate part of a printed board assembly that performs a circuit function — e.g., a resistor, a capacitor, or a transistor.

Discrete-wiring board - A base material upon which discrete-wiring techniques are used to obtain electrical interconnections.

Disturbed solder connection - A solder connection characterized by the appearance of motion between the metals being joined while the solder is solidifying.

Double-sided assembly - A packaging and interconnecting structure with components mounted on both the primary and secondary sides. (See also "Single-sided assembly.")

Double-sided base material - Base material with circuit patterns that have been electrically interconnected on both of its major surfaces that have been electrically interconnected.

Double-sided board - A printed board with a conductive pattern on both of its sides.

Drag soldering - Making soldered terminations by moving the solder side of a supported printed board with through-hole mounted components through the surface of a static pool of molten solder. (See also "Dip soldering.")

Dross - Oxide and other contaminants that form on the surface of molten solder.

Dual-in-line package (DIP) - A basically rectangular component package that has a row of leads extending from each of the longer sides of its body formed at right angles to a plane parallel to the base of its body.

Embedded component - A discrete component fabricated as an integral part of a printed board.

Eutectic (n.) - An alloy whose composition is indicated by the eutectic point on an equilibrium diagram or an alloy structure of intermixed solid constituents formed by a eutectic reaction.

Eutectic (v.) - An isothermal reversible reaction in which, on cooling, a liquid solution is converted into two or more intimately mixed solids, with the number of solids the same as the number of components in the system.

Excess solder connection - A solder connection characterized by the complete obscuring of the surfaces of the connected metals and/or by the presence of solder beyond the connection area.

Extraction tool - A device used for removing a contact from a connector body or insert, a component from a socket, or a printed board from its enclosure.

Face bonding - Attaching a die to a base material with its circuitry facing the base material.

Fatigue life - The number of cycles of stress that can be sustained prior to failure for a stated test condition.

Fatigue limit - The maximum stress below which a material can presumably endure an infinite number of stress cycles.

Fatigue strength - The maximum stress that can be sustained for a specified number of cycles without failure, with the stress being completely reversed within each cycle.

Fatigue-strength reduction factor (Kf) - The ratio of the fatigue strength of a member or specimen with no stress concentration to the fatigue strength with stress concentration.

Fault - Any condition that causes a device or circuit to fail to operate.

Fault simulation - A process that allows for the prediction or observation of a system's behavior in the presence of a specific fault, without actually having that fault occur.

Feature-based modeling - A computer-based modeling method based on the use of part features, rather than geometric entities.

Filler - A substance added to a material to improve its solidity, bulk, or other properties.

Fine-pitch technology (FPT) - A surface-mount assembly technology with component terminations on less than 0.625-mm (0.025-inch) centers.

Finger - A printed contact on or near any edge of a printed board used specifically for mating with an edge-board connector.

Finite-element analysis (FEA) - A computer-based analysis method that subdivides geometric entities into smaller elements and links a series of equations to each element so that they can be analyzed simultaneously.

Finite-element modeling (FEM) - The use of a model to represent a problem that can be evaluated by finite-element analysis.

Flag - The support area on a die or lead frame.

Flat pack - A rectangular component package with a row of leads extending from each of the longer sides of its body that are parallel to the base.

Flip chip - A leadless, monolithic, circuit element structure that electrically and mechanically interconnects to a base material through the use of conductive bumps.

Flip-chip mounting - The mounting and interconnecting of a flip-chip component to a base material.

Flow soldering - See "Wave soldering."

Flux - A chemically and physical active compound that, when heated, promotes the wetting by molten solder of a base metal surface by removing minor surface oxidation and other surface films and by protecting the surfaces from reoxidation during soldering.

Flux activity - The degree or efficiency with which a flux promotes wetting of a surface with molten solder. (See also "Solder-spread test" and "Wetting balance.")

Flux characterization - A series of tests that determines the basic corrosive and conductive properties of fluxes and flux residues.

Flux-cored solder - A wire or ribbon of solder that contains one or more continuous flux-filled cavities along its length.

Flux residue - A flux-related contaminant present on or near the surface of a solder connection.

Footprint - See "Land pattern."

Foreign material (soldering) - A lumpy, irregular coating that covers, or partially covers, particles of a different material located on the material or coating of the items to be soldered.

Fractional-factorial experiment - An experiment in which only a portion of the complete factorial is run.

Fusing - The combination of metals by means of melting, blending, and solidification.

Fusing flux - An activated organic fluid used in fusing a tin-lead plating on a basis metal. (The application of predominantly water-soluble fluids is usually followed by the use of a fusing oil.)

Fusing oil - A thermally stable, nonactivated fluid used in the fusing of tin-lead plating on a basis metal. (The application of predominantly water-soluble fluids is usually preceded by the use of a fusing flux.)

Gas blanket - A flowing inert gas atmosphere used to keep metallization from oxidizing.

Glass transition temperature - The temperature at which an amorphous polymer, or an amorphous region in a partially crystalline polymer, changes from being hard and relatively brittle to being viscous or rubbery.

Halide content - The ratio of the mass of free halides to the mass of solids in a flux, expressed as the mass percent of free chloride ions.

Hand soldering - Soldering using a soldering iron or other hand-held, operator-controllable apparatus.

Heat of fusion - The quantity of heat required to convert a unit weight of solid material to a liquid state.

High-quality/high-reliability soldering - A soldering technique in which the probability of obtaining a perfect fusing of metals, product cleanliness, and optimum electrical connectivity without damage to components or equipment has been statistically proven while using controlled processes, controlled environments, controlled facilities, approved applications, and trained certified personnel.

Hybrid circuit - An insulating base material with various combinations of interconnected film conductors, film components, semiconductor dice, passive components, and bonding wire that form an electronic circuit.

Icicle - See "Solder projection."

Illuminance - Luminous flux striking a surface.

Inorganic flux - An aqueous flux solution of inorganic acids and halides. (See also "Acid flux.")

Inspection facility - The combination of equipment, personnel, and procedure resources that perform inspection measurements and evaluations for the purpose of ascertaining the conformance of a product to applicable specifications.

Inspection lot - Units of a product identified and treated as a unique entity from which a sample is drawn and inspected to determine conformance with acceptability criteria.

Insufficient solder connection - A solder connection characterized by the incomplete coverage of one or more of the surfaces of the connected metals and/or by the presence of incomplete solder fillets.

Interfacial connection - A conductor that connects conductive patterns on both sides of a printed board — e.g., a plated through-hole. (See also "Interlayer connection.")

Interlayer connection - A conductor that connects conductive patterns on internal layers of a multilayer printed board — e.g., a plated through-hole. (See also "Interfacial connection.")

Intermetallic compound - An intermediate phase in an alloy system, with a

narrow range of homogeneity and relatively simple stoichiometric proportions, in which the nature of the atomic bonding can vary from metallic to ionic.
Interstitial via - See "Blind via" and "Buried via."

Junction temperature - The temperature of the region of a transition between p-type and n-type semiconductor material in a transistor or diode element.
Just-in-time (JIT) - Production control techniques that minimize inventory by delivering parts and material to a manufacturing facility just before they are incorporated into a product.

Land - A portion of a conductive pattern usually used for making electrical connections for component attachment.
Land pattern - A combination of lands used for mounting, interconnecting, and testing a particular component.
Landless hole - A plated through-hole without a land(s).
Layer-to-layer registration - The degree of conformity of a conductive pattern, or portion thereof, to that of any other conductive layer of a printed board.
Lead - A length of metallic conductor, insulated or uninsulated, used for electrical interconnections.
Leaded surface-mount component - A surface-mount component whose external connections consist of leads around and down the side of the package.
Lead extension - that part of a lead or wire that extends beyond a solder connection.
Lead frame - The metallic portion of a component package used to interconnect with semiconductor dice by wire bonding and to provide output terminal leads.
Lead mounting hole - See "Component hole."
Lead pin - See "Component pin."
Lead projection - The distance a component lead protrudes through the side of a printed board opposite the one upon which the component is mounted.
Lead wire - See "Component lead".
Leaded chip carrier - A chip carrier whose external connections consist of leads around and down the sides of the package. (See also "Leadless chip carrier.").
Leadless chip carrier - A chip carrier whose external connections consist of

metallized terminations that are an integral part of the component body. (See also "Leaded chip carrier.")

Leadless device - A chip component having no input or output leads.

Leadless inverted device - A shaped, metallized-ceramic form used as an intermediate carrier for diode or transistor dice which has been especially adapted for leadless surface mounting.

Leadless surface-mount component - A surface-mount component whose external connections consist of metallized terminations that are an integral part of the component body. (See also "Leaded surface-mount component.")

Manual soldering - See "Hand soldering."

Meniscus - The contour of a shape that results from the surface-tension forces that occur during wetting.

Metal migration - The electrolytic transfer of metal ions along an electrically conductive path from one metal surface to another when an electrical potential is applied to the two surfaces.

Microcircuit - A relatively high-density combination of equivalent circuit elements interconnected so as to perform as an indivisible electronic circuit component.

Microcircuit module - A combination of microcircuits, or of microcircuits and discrete components, interconnected so as to perform as an indivisible electronic circuit assembly.

Microcomponents - Small discrete components.

Microelectronics - The area of electronic technology with, or applied to, the realization of electronic systems from extremely small electronic elements, devices, or parts.

Mixed component-mounting technology - Component-mounting that uses both through-hole and surface-mounting technologies on the same packaging and interconnecting structure.

Module - A separable unit in a packaging scheme.

Monolithic integrated circuit - An integrated circuit in the form of a monolithic structure.

Mother board - A printed board assembly used for interconnecting arrays of plug-in electronic modules. (See also "Backplane.")

Multichip integrated circuit - See "Multichip module."

Multichip microcircuit - See "Multichip module."

Multichip module - A microcircuit module consisting primarily of closely spaced integrated circuit dice.
Multilayer printed board - The general term for a printed board that consists of rigid or flexible insulation materials and three or more alternate printed wiring and/or printed circuit layers that have been bonded together and electrically interconnected.
Multilayer printed circuit board - A multilayer printed board with two or more printed circuit layers.
Multilayer printed circuit board assembly - An assembly that uses a multilayer printed circuit board for component mounting and interconnecting purposes.
Multilayer printed wiring board - A multilayer printed board with only printed wiring for its conductive layers.
Multilayer printed wiring board assembly - An assembly that uses a multilayer printed wiring board for component mounting and interconnecting purposes.

Nominal-is-best characteristic - A parameter of quality that optimizes performance at its nominal value.
Nonactivated flux - A natural- or synthetic-resin flux without activators.
Nonpolar matter - A substance that cannot be dissolved in water and that is soluble in hydrophobic solvents.
Nonpolar solvent - A liquid that is not ionized to the extent that it is electrically conductive, that can dissolve nonpolar compounds (such as hydrocarbons and resins), and that cannot dissolve polar compounds (such as inorganic salts).
Nonwetting - The partial adherence of molten solder to a surface that it has contacted, with basis metal remaining exposed.
Normal distribution - A mathematically defined continuous distribution of values with a bell shape perfectly symmetrical about a mean value.

Outgassing - The gaseous emission from a printed board or component when a printed board assembly is exposed to a reduced pressure, heat, or both.

Package - The container for a circuit components used to protect its contents and to provide terminals for making connections to the rest of the circuit.
Packaging and interconnecting assembly - The general term for an as-

sembly with components mounted on either or both sides of a packaging and interconnecting structure.

Packaging density - The relative quantity of functions (components, interconnection devices, mechanical devices, etc.) per unit volume, usually expressed in qualitative terms, such as high, medium, and low.

Pad - See "Land."

Parallel-gap soldering - Passing an electrical current through a high-resistance space between two parallel electrodes to provide the energy required to make a soldered termination.

Partially clinched lead - A component lead inserted through a hole in a printed board and then formed to retain the component in place and to make metal-to-metal contact with a land prior to soldering. (See also "Clinched lead.")

Paste soldering - A soldering method that uses a solder paste applied to the land, device termination, or both.

Paste flux - A flux in the form of a paste that facilitates its application. (See also "Solder paste" and "Solder-paste flux.")

Pattern - The configuration of conductive and nonconductive materials on a base material, and the circuit configuration on related tools, drawings, and masters.

Perforated (pierced) solder terminal - A flat-metal solder terminal with an opening through which one or more wires is placed prior to soldering.

Photographic-reduction dimension - The dimensions on an artwork master, such as the distance between lines or between two specified points, that indicate the extent to which the artwork master is to be photographically reduced. (The value of the dimension refers to the 1-to-1 scale and must be specified.)

Pixel - The smallest definable picture element area capable of being displayed.

Plastic device - A semiconductor component for which the package or encapsulant is plastic.

Plated through-hole - A hole with plating on its walls that makes an electrical connection between conductive patterns on the internal or external layers, or both, of a printed board.

Plating - The chemical or electrochemical deposition of a metal on a surface.

Polar matter - A substance that can dissolve in water and hydrophilic solvents.

Polar solvent - A liquid ionized to the extent that it is electrically conductive, can dissolve nonpolar compounds (such as hydrocarbons and resins), and cannot dissolve polar compounds (such as inorganic salts).

Polymerized rosin - Rosin that has reacted with itself during the course of a soldering operation.

Porosity (solder) - A solder coating with an uneven surface and a spongy appearance that may contain a concentration of small pinholes and pits.

Positional tolerance - The amount that a feature is permitted to vary from its true-position location.

Postprocessing - Manipulating data after they have been generated or run through a batch process.

Postprocessor - A software procedure or program that interprets data and formats it into data readable by a numerically controlled machine or by other computer programs.

Preferred solder connection - A solder connection that is smooth, bright, and feathered-out to a thin edge to indicate proper solder flow and wetting action, to cover all bare metal within the solder connection and to eliminate sharp protrusions of solder or contamination (e.g., embedded foreign material).

Preheating - Raising the temperature of a material(s) above the ambient temperature to reduce thermal shock and to influence the dwell time during subsequent elevated-temperature processing.

Pretinning - See "tinning."

Primary side - The side of a packaging and interconnecting structure so defined on the master drawing, usually the side that contains the most complex or numerous components.

Printed board - The general term for completely processed printed circuit and printed wiring configurations, including single-sided, double-sided, and multilayer boards with rigid, flexible, and rigid-flex base materials.

Printed board assembly - The generic term for an assembly that uses a printed board for component mounting and interconnecting purposes.

Printed circuit - A conductive pattern composed of printed components, printed wiring, or a combination thereof, formed in a predetermined arrangement on a common base. (Also a generic term used to describe a printed board produced by any of a number of techniques.)

Printed circuit board - A printed board that provides both point-to-point connections and printed components in a predetermined arrangement on a common base. (See also "Printed wiring board.")

Printed circuit board assembly - An assembly that uses a printed wiring board for component mounting and interconnecting purposes.

Printed component - A part (such as an inductor, resistor, capacitor, or transmission line) formed as part of the conductive pattern of a printed board.

Printed wiring - A conductive pattern that provides point-to-point connections, but not printed components, in a predetermined arrangement on a common base. (See also "Printed circuit.")

Printed wiring board - A printed board that provides point-to-point connections, but not printed components, in a predetermined arrangement on a common base. (See also "Printed circuit board".)

Printed wiring board assembly - An assembly that uses a printed wiring board for component mounting and interconnecting purposes.

Production board - A printed board or discrete-wiring board manufactured in accordance with the applicable detailed drawings, specifications, and procurement requirements.

Pull strength - See "Bond strength."

Pulse soldering - Soldering by the heat generated by pulsing an electrical current through a high-resistance point of the joint area and the solder.

Push-off strength - The amount of force required to dislodge a leadless component by the application of a force parallel to the surface upon which it is mounted.

Qualification agency - The organization used to perform documentation reviews and audits of an inspection or testing facility.

Qualification testing - The demonstration of the ability to meet all of the requirements specified for a product.

Quality-conformance test circuity - A portion of a printed board panel that contains a complete set of test coupons used to determine the acceptability of the board(s) on the panel.

Quality-conformance testing - Qualification testing performed on a regularly scheduled basis to demonstrate the continued ability of a product to meet all the quality requirements specified.

Randomization - The random selection of experimental runs to minimize biases due to unknown or uncontrollable factors in an experimental design.

Random-effects model - A specific experimental treatment whereby a random sample is taken from a larger population in such a manner that the conclusions reached can be extended to the entire population and the inferences are not restricted to the experimental levels.
Random sample - A set of individual items taken from a population in such a way that each possible individual has an equal chance of being selected.
Reflow soldering - The joining of mating surfaces that have been tinned and/or soldered, placing them together, heating them until the solder fuses, and allowing them to cool in the joined position.
Registration - The degree of conformity of the position of a pattern (or portion thereof), hole, or other feature to its intended position on a product.
Reliability - The lifetime performance characteristics of a device or assembly.
Repair(ing) - The act of restoring the functional capability of a defective article in a manner that precludes compliance of the article with applicable drawings or specifications.
Residue - Any visual or measurable form of process-related contamination.
Resin - A natural or synthetic resinous material. (See also "Rosin" and "Synthetic resin.")
Resin flux - A resin and small amounts of organic activators in an organic solvent.
Resistance soldering - Soldering by a combination of pressure and heat generated by passing a high current through two mechanically joined conductors.
Rework(ing) - The act of reprocessing noncomplying articles through the use of original or alternate equivalent processing, in a manner that ensures compliance of the article with applicable drawings or specifications.
Rosin - A hard, natural resin consisting of abietic and primaric acids and their isomers, some fatty acids, and terepene hydrocarbons, extracted from pine trees and subsequently refined.
Rosin flux - Rosin in an organic solvent or rosin as a paste with activators.

Saponifier - An aqueous organic or inorganic-base solution with additives that promote the removal of rosin and/or water-soluble flux.
Screen printing - Transferring an image to a surface by forcing a suitable medium through an imaged-screen mesh with a squeegee.
Scrubbing action - Rubbing the lead wire and bonding land to break up oxide layers and to improve bondability.

Secondary side - The side of a packaging and interconnecting structure opposite the primary side and equivalent to the solder side in through-hole mounting technology.

Seed layer - See "Activating."

Semiconductor - A solid material, such as silicon, that has a resistivity midway between that of a conductor and that of a resistor.

Separable component part - A replaceable component part with a body not chemically bonded to the base material (excluding protective coatings, solder, and potting compounds).

Shear strength - The force in Newtons required to shear apart adhesive-bonded (and cured) materials and/or components.

Sheet-metal contact - A type of connector contact consisting of flat spring metal formed by either stamping or bending.

Shield - The material around a conductor or group of conductors that limits electromagnetic and/or electrostatic interference.

Sigma - The lowercase Greek letter used to designate a standard deviation of a population.

Simulated aging - The artificial exposure of material to conditions of both high and low temperature and humidity to produce changes in its properties that are similar to those that occur during extended exposure to normal environmental conditions.

Single-in-line package (SIP) - A component package with one straight row of pins or wire leads.

Single-sided assembly - A packaging and interconnecting structure with components mounted on only one side. (See also "Double-sided assembly.")

Single-sided board - A printed board with a conductive pattern on only one side.

Slump - The distance a substance (e.g., adhesive) moves after it has been applied and cured.

Smeared bond - A bond impression that has been distorted or enlarged by excess lateral movement of the bonding tool or holding device fixture.

Solder - A metal alloy with a melting temperature below 427°C (800°F).

Solderability - The ability of a metal to be wetted by molten solder.

Solder bridging - The unwanted formation of a conductive path of solder between conductors.

Solder coat - A layer of solder applied directly from a molten solder bath to a conductive pattern.

Solder connection - An electrical/mechanical connection that employs solder to join two or more metal surfaces. (See also "Cold solder connection," "Disturbed solder connection," "Excess solder connection," "Insufficient solder connection," "Preferred solder connection," and "Solder connection pinhole.")

Solder connection pinhole - A small hole that penetrates from the surface of a solder connection to a void of indeterminate size within the solder connection.

Solder contact - A type of connector contact whose non-mating end is a hollow cylinder, cup, eyelet or hook that can be soldered to a wire in contact with it.

Solder cream - See "Solder paste."

Solder embrittlement - The reduction in mechanical properties of a metal as a result of local penetration of solder along grain boundaries.

Solder fillet - A normally concave surface of solder at the intersection of the metal surfaces of a solder connection.

Solder levelling - A solder-coating process in which a heated gas or other medium levels and removes excess solder.

Solder mask - See "Solder resist."

Solder paste - Finely divided particle of solder, with additives to promote wetting and to control viscosity, tackiness, slumping, drying rate, and so on, that is suspended in a cream flux.

Solder-paste flux - Solder paste without solder particles.

Solder plug - A core of solder in a plated through-hole.

Solder projection - An undesirable protrusion of solder from a solidified solder joint or coating.

Solder resist - A resist that provides protection from the action of solder.

Solder side - The secondary side of a single-sided assembly.

Solder spatter - Extraneous fragments of solder with an irregular shape.

Solder-spread test - The determination of a relative measure of solder flux efficiency that is obtained by calculating the area of spread of a specified weight of solder placed on a specially prepared and fluxed metallic surface.

Solder terminal - An electrical/mechanical connection device used to terminate a discrete wire or wires by soldering. (See also "Cup solder terminal," "Perforated (pierced) solder terminal," and "Turret solder terminal.")

Solder webbing - A continuous film or curtain of solder parallel to, but not necessarily adhering to, a surface that should be free of solder.

Solder wicking - The capillary movement of solder between metal surfaces, such as strands of wire.

Soldering - Joining metallic surfaces with solder and without melting the base material.

Soldering ability - The ability of a specific combination of components to facilitate the formation of a proper solder joint.

Soldering flux - See "Flux".

Soldering iron tip - The portion of a soldering iron used for the application of heat that melts the solder.

Soldering oil (blanket) - Liquid formulations used in intermix wave soldering and as coverings on static and wave soldering pots to eliminate dross and to reduce surface tension during soldering.

Solvent cleaning - The removal of organic and inorganic soils, using a blend of polar and nonpolar organic solvents.

Standard deviation of a population - A measure of the distribution of a population about a mean value equal to the square root of the variance of a process output. (See also "Sigma.")

Standard deviation of a sample - A measure of the distribution of test data about a mean value equal to the square root of the variance of a sampling characteristic. (See also "Sigma.")

Standoff solder terminal - See "Turret solder terminal."

Statistical control - The condition of describing a process from which all special causes of variation have been eliminated and from which only common causes remain.

Statistical hypothesis - An assumption about a population being sampled.

Statistical process control (SPC) - The use of statistical techniques to analyze a process or its output to enable appropriate action to achieve and maintain a state of statistical control and to improve process capability.

Statistical quality control (SQC) - The use of statistical techniques to document and assure compliance with requirements.

Step soldering - Making solder connections by sequentially using solder alloys with successively lower melting temperatures.

Straight-through lead - A component lead that extends through a hole and is terminated without subsequent forming.

Stress relief - The portion of a component lead or wire lead formed in such a way as to minimize mechanical stresses after the lead is terminated.

Surface-mount component (device) - A leaded or leadless component (device) capable of being attached to a printed board by surface mounting.
Stud-mount termination - See "Straight-through lead."
Surface mounting - The electrical connection of components to the surface of a conductive pattern that does not utilize component holes.
Surface tension - The natural, inward, molecular-attraction force that spreads a liquid at its interface with a solid material.
Swaged lead - A component lead wire that extends through a hole in a printed board, its lead extension flattened (swaged) to secure the component to the board during manufacturing operations.
Synthetic activated flux - A highly activated organic flux whose post-soldering residues are soluble in halogenated solvents.
Synthetic resin - A synthetic organic material or a chemically-treated natural resin miscible in water.

Tab - See "Printed contact."
TAB - See "Tape automated bonding."
Tape automated bonding - A fine-pitch technology that provides interconnections between die and base materials with conductors that are on a carrier tape.
Tape - See "Carrier tape."
Terminal - A metalic device used for making electrical connections. (See also "Solder terminal.")
Terminal hole - See "Component hole."
Thermal coefficient of expansion (TCE) - See "Coefficient of thermal expansion (CTE)".
Thermal conductivity - The property of a material that describes the rate at which heat is conducted through a unit area of the material for a given driving force.
Thermal expansion mismatch - The absolute difference between the thermal expansion of two components or materials. (See also "Coefficient of thermal expansion (CTE)")
Thermocompression bonding - Joining two materials without an intermediate material by the application of pressure and heat in the absence of electrical current.
Thermosonic bonding - Terminations made by combining thermocompres-

sion and ultrasonic bonding principles.

Thixotropy - A property of a substance (e.g., an adhesive system) that allows it to get thinner upon agitation and thicker upon subsequent rest.

Through-hole mounting - Electrically connecting components to a conductive pattern by the use of component holes.

Tinning - The application of molten solder to a basis metal to increase its solderability.

Tolerance - The total amount by which a specific dimension is permitted to vary.

Touch-up - See "Reworking".

Transfer soldering - The use of a soldering iron to transfer a measured amount of solder, in the form of a ball, chip, or disc, to a solder connection.

Turret solder terminal - A round post-type stud (standoff) solder terminal with a groove or grooves around which one or more wires are wrapped prior to soldering.

Two-sided board - See "Double-sided board."

Ultrasonic bonding - A termination process that uses ultrasonic frequency vibration energy and pressure to make the joint.

Ultrasonic soldering - Fluxless soldering wherein molten solder is vibrated at ultrasonic frequencies while the joint is made.

Unsupported hole - A hole in a printed board that does not contain plating or another type of conductive reinforcement.

Vapor-phase soldering - A reflow soldering method based on the exposure of the parts to be soldered to hot vapors of a liquid with a boiling point sufficiently high to melt the solder used.

Via - A plated through-hole used as an interlayer connection, but which is not intended to accommodate a component lead or other reinforcing material. (See also "Blind via" and "Buried via.")

Void - The absence of any substances in a localized area.

Wave soldering - A process wherein an assembled printed board is brought in contact with the surface of a continuously flowing and circulating mass of solder.

Water-soluble organic flux - An organic chemical soldering flux soluble in water.

Wedge bond - A wire bond made with a wedge tool.

Wedge tool - A bonding tool in the general form of a wedge, with or without a guide hole to position wire under its bonding face.

Wetting - The formation of a relatively uniform, smooth, unbroken, and adherent film of solder to a basis metal.

Wetting balance - An instrument used to measure wetting performance and solderability.

Whisker - A slender acicular (needle-shaped) metallic growth between a conductor and a land.

Wicking - The capillary absorption of a liquid along the fibers of a base material. (See also "Solder wicking.")

Wipe soldering - Forming a joint by applying semifluid solder and shaping the joint by rubbing with a greased cloth pad.

Wire lead - A length of uninsulated solid or stranded wire readily formable to a desired configuration and used for electrical interconnections. (See also "Component lead.")

Index